AF581712

Visual Servoing in Robotics

Visual Servoing in Robotics

Editor

Jorge Pomares

MDPI • Basel • Beijing • Wuhan • Barcelona • Belgrade • Manchester • Tokyo • Cluj • Tianjin

Editor
Jorge Pomares
Department of Physics,
Systems Engineering and Signal Theory,
University of Alicante
Spain

Editorial Office
MDPI
St. Alban-Anlage 66
4052 Basel, Switzerland

This is a reprint of articles from the Special Issue published online in the open access journal *Electronics* (ISSN 2079-9292) (available at: https://www.mdpi.com/journal/electronics/special_issues/Visual_Servoing).

For citation purposes, cite each article independently as indicated on the article page online and as indicated below:

LastName, A.A.; LastName, B.B.; LastName, C.C. Article Title. *Journal Name* **Year**, *Volume Number*, Page Range.

ISBN 978-3-0365-0344-8 (Hbk)
ISBN 978-3-0365-0345-5 (PDF)

Cover image courtesy of Jorge Pomares.

Contents

About the Editor

Jorge Pomares (Prof. Dr.) obtained his degree in Computer Engineering and his PhD at the University of Alicante. He has been a member of the Department of Physics, Systems Engineering and Signal Theory at the University of Alicante since 2003. Since December 2017, he has been Full Professor in the aforementioned department in the area of systems engineering and automatics, being founder of the Human Robotics research group. His research career has focused on the field of robotics, visual servoing for guiding robots, space robotics, and robot control and manipulation. Within these fields, Prof. Dr. Pomares has participated in more than 15 research projects. His projects have covered robot guidance by vision, visual servoing, and the control of assistive robotics. These projects have allowed not only the development of robot guidance strategies but also the design of new approaches for the control of robotic systems. In addition, he has worked on the design of new strategies for guiding robots with different characteristics: robot manipulators, robotic hands, mobile robots, mobile manipulators, space robots, exoskeletons, and so on. Within this field, he has authored more than 40 JCR articles and presented over 100 contributions at national and international conferences. Throughout his research career, Prof. Dr. Pomares has collaborated with more than ten national and international research groups in the field of robotics, robot control systems, and space robotics. Most notably, regarding collaborations with foreign research groups, during the last few years, he has been collaborating with the Institute of Industrial Systems of Greece, the University of Lulea, and the Centre for Autonomous and Cyberphysical Systems at the University of Cranfield. These collaborations have been focused on the development of new methods of nonlinear control and filtering for robotic applications. This research aims to solve control, estimation, and filtering problems in advanced models of robotic manipulators and vehicles. The methods which have been developed are generic and applicable to a wide range of robotic systems. These control methods apply to to the following types of robots: multi-DOF rigid-link robots, manipulators subject to input/output delays, underactuated robots and redundant manipulators, closed-chain robotic systems, exoskeletons, flexible-link robots, and space robotics. Additionally, Prof. Dr. Pomares is collaborating with the Centre for Autonomous and Cyberphysical Systems at the University of Cranfield, within the research field of space robotics.

Preface to "Visual Servoing in Robotics"

Today, computer vision systems are integrated into a wide range of industrial applications as a means of detecting defects in production lines, of object/pattern recognition, and as a surface scanning method, among other uses. Within these industrial applications, it is important to highlight visual servoing systems for the guidance of robotic manipulators. The main goal of visual servoing systems is the use of computer vision as a method for the guidance of robots. These control systems are fed with the information acquired by one or several cameras; this controller then determines the control actions to move the manipulator in order to carry out its task. Visual servoing approaches can be applied for the guidance of not only robot manipulators but also different kinds of robots, such as mobile robots, aerial robots, or parallel robots. This book includes different areas of research where visual servoing approaches are applied for the guidance of mobile manipulators, parallel robots, or even teleoperated robots.

As is evident throughout the book, visual servoing systems are widespread and extensively applied, not only in research laboratories but also in a wide range of additional applications, from industrial robotics to service robotics. This book also includes papers that describe new and interesting applications of visual servoing systems. One notable application is the use of visual servoing in apple-picking robots. The image-based visual servo control method is adopted to control the manipulator in order to improve the robot's grasping accuracy during the picking process. Additionally, a spatial trajectory optimization method for a spray-painting robot is presented in this book.

One factor which has favored the growth of visual servoing systems, their diffusion, and the extension of their application is the increase in the capture and processing capabilities of today's cameras (such as the RGBD cameras presented in this book), as well as the equipment used for processing data. This has allowed for the computation of vast amounts of information in less time, avoiding possible delays and helping to guide the robot in a smooth manner. Despite the efforts of the last decade dedicated to improving different aspects of the development of these visual servoing systems, even nowadays, there is considerable ongoing research to investigate means of increasing the robustness of such systems. Consequently, this book includes an image feature reconstruction algorithm based on the Kalman filter to handle feature loss during tracking.

Jorge Pomares
Editor

 electronics

Editorial

Visual Servoing in Robotics

Jorge Pomares

Department of Physics, Systems Engineering and Signal Theory, University of Alicante, 03690 Alicante, Spain; jpomares@ua.es

Received: 4 October 2019; Accepted: 1 November 2019; Published: 6 November 2019

1. Introduction

Visual servoing is a well-known approach to guide robots using visual information. Image processing, robotics and control theory are combined in order to control the motion of a robot depending on the visual information extracted from the images captures by one or several cameras. With respect to vision issues, different problems are currently under research such as the use of different kinds of image features (or different kinds of cameras), image processing at high velocity, convergence properties, etc. Furthermore, the use of new control schemes allows the system to behave more robustly, efficiently, or compliantly with less delays. Related issues such as optimal and robust approaches, direct control, path tracking or sensor fusion allows the application of the visual servoing systems in different domains.

In so-called image-based visual servoing systems, the control law is calculated using directly visual information. These last systems do not need a complete 3D reconstruction of the environment. For tasks that require high precision, speed or response, several works suggest that it may be beneficial to take into account the dynamics of the robot when designing visual servoing control laws. The type of visual control systems that consider the dynamics of the robot in the control law are often referred to as direct or dynamic visual control systems. However, for simplicity, indirect visual servoing schemes are mostly used in the literature.

Nowadays, the application fields of the visual servoing systems are very wide, and include research and application fields such as navigation and localization of mobile robots, guidance of humanoid robots, robust and optimal control of robots, manipulation, intelligent transportation, deep learning and machine learning in visual servoing and visual guidance of field robotics (aerial robots, assistive Robots, medical robots, etc.).

2. The Present Issue

This special issue consists of eight papers covering important topics in the field of visual servoing. In [1], an enhanced switch image-based visual servoing system for a 6 degrees of freedom industrial robot is proposed. An image feature reconstruction algorithm based on the Kalman filter is presented to handle feature loss during the tracking. Visual servoing approaches can be applied for the guidance of different kinds of robots such as mobile robots, aerial robots or parallel robots. The latter is the case described in [2], where an optical coordinate measuring machine is employed to stabilize parallel robots. In this case, the dynamic model parameters are identified by using a non-linear optimisation technique. In [3], a direct image-based visual servoing system is used for the guidance of a mobile manipulator. This approach considers not only kinematic properties of the robot, but also dynamic ones for guiding both the robot base and the manipulator arm. An optimal control approach is used in this paper.

This special issue also includes papers that describe new and interesting applications of the visual servoing systems such as the ones described in [4] or [5]. In [4], a visual servoing system is applied to an apple-picking robot. The image-based visual servo control method is adopted to control the manipulator in order to improve the grasping accuracy in the picking process. The joint control performance of the control system has been improved by the proposed adaptive fuzzy neural network sliding-mode control algorithm. Additionally, in [5], a spatial trajectory optimization method of a spray-painting robot is proposed.

Additionally, visual servoing approaches are very related with the necessity of estimating parameters or variables used in the control process. For example, in [6], a visual-based method to estimate robot orientation with RGB-D cameras is proposed. In [7], the reaction force of the end effector and second link of a three-degree of freedom hydraulic servo system with master–slave manipulators sliding mode control is determined with a sliding perturbation observer. Also, bilateral control is used to estimate the reaction force of the master device which is provided to the operator to handle the master device. Finally, in [8] a novel measurement system to visualize the motion-to-photon latency with time-series data in real time is proposed.

3. Future

While visual servoing systems have been an important field of research in the last years, several major challenges still remain. Tasks such as tracking, positioning, detection, segmentation, and localization play a critical role in visual servoing and different research is currently ongoing to increase the robustness of visual controllers. Additionally, new control approaches such as optimal control, robust control, dynamic control or predictive control will provide this kind of system with new dynamic properties. Furthermore, new computer vision systems, electronics, computers, etc., will offer new and interesting capabilities for the application of visual servoing in new kinds of robotics systems such as autonomous driving cars, humanoid robots, aerial robots, service robotics, UAV, parallel robots, space robotics, etc.

Acknowledgments: First of all, we would like to thank all researchers who submitted articles to this special issue for their excellent contributions. We are also grateful to all reviewers who helped in the evaluation of the manuscripts and made very valuable suggestions to improve the quality of contributions. We would like to acknowledge the editorial board of Electronics, who invited us to guest edit this special issue. We are also grateful to the Electronics Editorial Office staff who worked thoroughly to maintain the rigorous peer-review schedule and timely publication.

Conflicts of Interest: The author declares no conflict of interest.

References

1. Ghasemi, A.; Li, P.; Xie, W.-F.; Tian, W. Enhanced Switch Image-Based Visual Servoing Dealing with FeaturesLoss. *Electronics* **2019**, *8*, 903. [CrossRef]
2. Li, P.; Ghasemi, A.; Xie, W.; Tian, W. Visual Closed-Loop Dynamic Model Identification of Parallel Robots Based on Optical CMM Sensor. *Electronics* **2019**, *8*, 836. [CrossRef]
3. Belmonte, Á.; Ramón, J.L.; Pomares, J.; Garcia, G.J.; Jara, C.A. Optimal Image-Based Guidance of Mobile Manipulators using Direct Visual Servoing. *Electronics* **2019**, *8*, 374. [CrossRef]
4. Chen, W.; Xu, T.; Liu, J.; Wang, M.; Zhao, D. Picking Robot Visual Servo Control Based on Modified Fuzzy Neural Network Sliding Mode Algorithms. *Electronics* **2019**, *8*, 605. [CrossRef]
5. Chen, W.; Wang, X.; Liu, H.; Tang, Y.; Liu, J. Optimized Combination of Spray Painting Trajectory on 3D Entities. *Electronics* **2019**, *8*, 74. [CrossRef]
6. Guo, R.; Peng, K.; Zhou, D.; Liu, Y. Robust Visual Compass Using Hybrid Features for Indoor Environments. *Electronics* **2019**, *8*, 220. [CrossRef]
7. Kallu, K.D.; Wang, J.; Abbasi, S.J.; Lee, M.C. Estimated Reaction Force-Based Bilateral Control between 3DOF Master and Hydraulic Slave Manipulators for Dismantlement. *Electronics* **2018**, *7*, 256. [CrossRef]
8. Choi, S.-W.; Lee, S.; Seo, M.-W.; Kang, S.-J. Time Sequential Motion-to-Photon Latency Measurement System for Virtual Reality Head-Mounted Displays. *Electronics* **2018**, *7*, 171. [CrossRef]

Article

Enhanced Switch Image-Based Visual Servoing Dealing with Features Loss

Ahmad Ghasemi [1], Pengcheng Li [1], Wen-Fang Xie [1,*] and Wei Tian [2]

[1] Department of Mechanical, Industrial & Aerospace Engineering, Concordia University, Montreal, QC 1455, Canada
[2] College of Mechanical and Electrical Engineering, Nanjing University of Aeronautics and Astronautics, Nanjing 210016, China
* Correspondence: wfxie@encs.concordia.ca

Received: 18 July 2019; Accepted: 9 August 2019; Published: 15 August 2019

Abstract: In this paper, an enhanced switch image-based visual servoing controller for a six-degree-of-freedom (DOF) robot with a monocular eye-in-hand camera configuration is presented. The switch control algorithm separates the rotating and translational camera motions and divides the image-based visual servoing (IBVS) control into three distinct stages with different gains. In the proposed method, an image feature reconstruction algorithm based on the Kalman filter is proposed to handle the situation where the image features go outside the camera's field of view (FOV). The combination of the switch controller and the feature reconstruction algorithm improves the system response speed and tracking performance of IBVS, while ensuring the success of servoing in the case of the feature loss. Extensive simulation and experimental tests are carried out on a 6-DOF robot to verify the effectiveness of the proposed method.

Keywords: image-based visual servoing; image feature loss; industrial robots; switch control

1. Introduction

Visual servoing has been employed to increase the deftness and intelligence of industrial robots, especially in unstructured environments [1–4]. Based on how the image data are used to control the robot, visual servoing is classified into two categories: position-based visual servoing (PBVS) and image-based visual servoing (IBVS). A comprehensive analysis of the advantages and drawbacks of the aforementioned methods can be found in [5]. This paper focuses on addressing some issues in IBVS.

Many studies have been conducted to overcome the weaknesses of IBVS and improve its efficiency [6–9]. However, the performance of most reported IBVS is not sufficiently high to meet the requirements of industrial applications [10]. An efficient IBVS feasible for practical robotic operations requires a fast response with strong robustness to feature loss. One obvious way to increase the speed of IBVS is to increase the gain values in the control law. However, there is a limitation on the application of this strategy because the high gain in the IBVS controller tends to create shakiness and instability in the robotic system [11]. Moreover, the stability of the traditional IBVS system is proven only in an area around the desired position [5,12]. Furthermore, when the initial feature configuration is distant from the desired one, the converging time is long, and possible image singularities may lead to IBVS failure. To address this issue, a switching scheme is proposed to switch the control signal between low-level visual servo controllers, i.e., homography-based controller [13] and affine-approximation controller [14].

In our previous work [15,16], the idea of switch control in IBVS was proposed to switch the controller between end-effector's rotating and translational movements. Although it has been demonstrated that the switch control can improve the speed and tracking performance of IBVS and avoid some of its inherent drawbacks, feature loss caused by the camera's limited FOV still prevents the method from being fully efficient and being applicable to real industrial robots.

The visual features contain much information such as the robots' pose information, the tasks' states, the influence of the environment, the disturbance to the robots, etc. The features are directly related to the motion screw of the end-effector of the robot. The completeness of the feature set during visual servoing is key to fulfilling the task successfully. Many features have been used in visual servoing such as feature points, image moments, lines, etc. The feature points are known for the ease of image processing and extraction. It is shown that at least three image points are needed for controlling a 6-DOF robot [17]. Hence, four image points are usually used for visual servoing. However, the feature points tend to leave the FOV during the process of visual servoing. A strategy is needed to handle the situation where the features are lost.

There are two main approaches to handle feature loss and/or occlusion caused by the limited FOV of the camera [18]. In the first approach, the controller is designed to avoid occlusion or feature loss, while in the second one, the controller is designed to handle the feature loss.

In the first approach, several techniques have been developed to avoid the feature loss or occlusion. In [19], occlusion avoidance was considered as the second task besides the primary visual servoing task. In [20], a reactive unified convex optimization-based controller was designed to avoid occlusion during teleoperation of a dual-arm robot. Some studies have been carried out in visual trajectory planning considering feature loss avoidance [21–23]. Model predictive control methods have been adopted in visual servoing to prevent feature loss due to its ability to deal with constraints [24–28]. In [29], predictive control was employed to handle visibility, workspace, and actuator constraints. Despite the success of the studies on preventing feature loss, they suffered from the limited maneuvering workspace of the robot, due to the conservative design required to satisfy many constraints.

In the second approach, the controller tries to handle the feature loss instead of avoiding it. When the loss or occlusion of features occurs, if the remaining visible features are sufficient to generate the non-singular inverse of the image Jacobian matrix, the visual servoing task can still be carried out successfully. In this situation, the rank of the relative Jacobian matrix must be the same as the degrees of freedom [30]. However, this method is no longer effective when the number of remaining visible features become too small to guarantee the full-rankness of the image Jacobian matrix. As studied in [31], another solution is to foresee the position of the lost features and to continue the control process using the predicted features until they become visible again. This method allows partial or complete loss or occlusion of the features. In the second approach [30,31], the classical IBVS control is employed as the control method, which does not usually provide a fast response.

In this paper, an enhanced switch image-based visual servoing (ESIBVS) method is presented in which a Kalman filter-based feature prediction algorithm is proposed and is combined with our previous work [15,16] to make the switch IBVS control robust in reaction to feature loss. The feature prediction algorithm can predict the lost feature points based on the previously-estimated points. The switch control with the improved tracking performance along with the robustness to feature loss makes it more feasible for industrial robotic applications. To validate the proposed controller, extensive simulations and experiments have been conducted on a 6-DOF Denso robot with a monocular eye-in-hand vision system.

The structure of the paper is given as follows. Section 2 gives a description of the problem. In Section 3, the feature reconstruction algorithm is presented. In Section 4, the controller design algorithm is developed. In Section 5, the simulation results are given. Experimental results are presented in Section 6, and finally, the concluding remarks are given in Section 7.

2. Problem Statement

In IBVS, the object with (X, Y, Z) coordinates with respect to camera has the projected image coordinates (x, y) in the camera image (Figure 1). The feature's positions and the desired ones for the nth feature in the image plane can be denoted by:

$$s_n = [x_n \; y_n]^T, \qquad s_{dn} = [x_{dn} \; y_{dn}]^T \tag{1}$$

Thus, the vector of s and s_d is defined as:

$$s = \begin{bmatrix} s_1 \\ \vdots \\ s_n \end{bmatrix} = \begin{bmatrix} x_1 \\ y_1 \\ \vdots \\ x_n \\ y_n \end{bmatrix}, \; s_d = \begin{bmatrix} s_{d1} \\ \vdots \\ s_{dn} \end{bmatrix} = \begin{bmatrix} x_{d1} \\ y_{d1} \\ \vdots \\ x_{dn} \\ y_{dn} \end{bmatrix}. \tag{2}$$

The goal of the IBVS task is to generate camera velocity commands such that the actual features and the desired ones are matched in the image plane. The velocity of the camera is defined as $V_c(t)$. The camera and image feature velocities are related by:

$$\dot{s} = J_{img} V_c, \tag{3}$$

where,

$$J_{img} = \begin{bmatrix} J_{img}(s_1, Z_1) \\ \vdots \\ J_{img}(s_n, Z_n) \end{bmatrix}, \tag{4}$$

which is called the image Jacobian matrix and $Z_1, \ldots, Z_n$ are the depths of the features $s_1, ..., s_n$. In this study, the system configuration is set as eye-in-hand, and the number of features is $n = 4$. Furthermore, it is assumed that all the features share the same depth Z. Considering these assumptions, the image Jacobian matrix for the nth feature is given in [17]:

$$J_{img}(s_n) = \begin{bmatrix} \frac{f}{Z} & 0 & -\frac{x_n}{Z} & -\frac{x_n y_n}{f} & \frac{f^2+x_n^2}{f} & -y_n \\ 0 & \frac{f}{Z} & -\frac{y_n}{Z} & \frac{-f^2-y_n^2}{f} & \frac{x_n y_n}{f} & x_n \end{bmatrix}, \tag{5}$$

where f is the focal length of the camera.

The velocity of the camera can be calculated by manipulating (3):

$$V_c = \overset{+}{J}_{img} \dot{s}, \tag{6}$$

where $\overset{+}{J}_{img}$ is the pseudo-inverse of the image Jacobian matrix. The error signal is defined as $e = s - s_d$. If we let $\dot{e} = -K_a e$, the traditional IBVS control law may be designed as:

$$V_c = -K_a \overset{+}{J}_{img} e, \tag{7}$$

where K_a is the proportional gain.

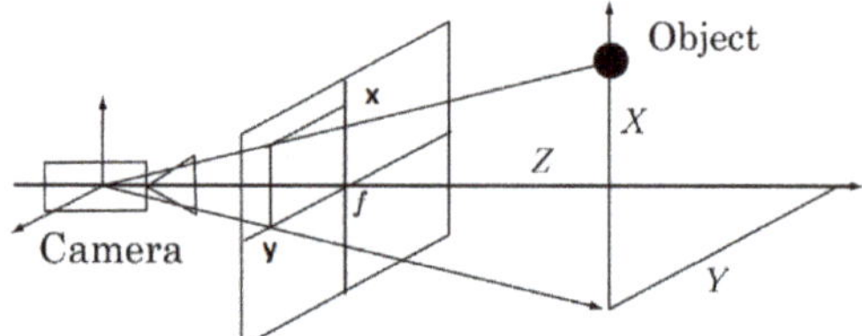

Figure 1. Schematic of the camera model.

While guiding the robot end-effector to make the desired image features match the actual ones, some unexpected situations may occur in IBVS. The first case is feature loss: i.e., some or all of the image features may go beyond the camera's FOV (Figure 2). The second case is feature occlusion: i.e., some or all of the image features temporarily become invisible to the camera due to obstacles. The goal of this paper is to improve the performance of IBVS in terms of response time and tracking performance, while dealing with the feature loss situation. To reach this goal, the performance of the switch method in our previous work [15,16] is enhanced when it is combined with the proposed feature reconstruction algorithm.

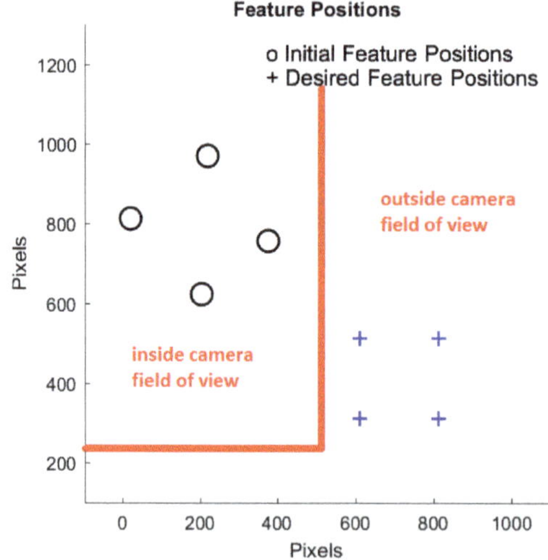

Figure 2. Desired and initial feature positions inside and outside the camera's field of view.

3. Feature Reconstruction Algorithm

The velocity of the camera $V_c \in \mathbb{R}^{(6\times1)}$ can be divided into the translational velocity $V \in \mathbb{R}^{(3\times1)}$ and rotating velocity $\omega \in \mathbb{R}^{(3\times1)}$. Therefore, it can be expressed as:

$$V_C = \begin{bmatrix} V \\ \omega \end{bmatrix} = \begin{bmatrix} V_x \\ V_y \\ V_z \\ \omega_x \\ \omega_y \\ \omega_z \end{bmatrix}, \tag{8}$$

Furthermore, for the nth feature $(n = 1, 2, .., 4)$, the image Jacobian matrix in (5) can be divided into the translational part $J_t(s_n)$ and the rotating part $J_r(s_n)$:

$$J_{img}(s_n) = \begin{bmatrix} J_t(s_n) & J_r(s_n) \end{bmatrix}, \tag{9}$$

where,

$$J_t(s_n) = \begin{bmatrix} \frac{f}{Z} & 0 & -\frac{x_n}{Z} \\ 0 & \frac{f}{Z} & -\frac{y_n}{Z} \end{bmatrix} \tag{10}$$

and:

$$J_r(s_n) = \begin{bmatrix} -\frac{x_n y_n}{f} & \frac{f^2+x_n^2}{f} & -y_n \\ \frac{-f^2-y_n^2}{f} & \frac{x_n y_n}{f} & x_n \end{bmatrix}, \tag{11}$$

where x_n and y_n are the feature coordinates in the image space.

In the design of the switch controller, the movement of the camera during the control task is divided into three different stages [15,16]. In the first stage, the camera has only pure rotation. In the second stage, the camera has only translational movement. Finally, in the third stage, both camera rotation and translation are used to carry out the fine-tuning.

Considering (3), (8), (10) and (11), the feature velocity in the image plane can be expressed as:

In the pure translational stage (first stage):

$$\begin{cases} \dot{x}_n = \frac{f}{Z}V_x - \frac{x_n}{Z}V_z \\ \dot{y}_n = \frac{f}{Z}V_y - \frac{y_n}{Z}V_z. \end{cases} \tag{12}$$

In the pure rotating stage (second stage):

$$\begin{cases} \dot{x}_n = -\frac{x_n y_n}{f}\omega_x + \frac{f^2+x_n^2}{f}\omega_y - y_n\omega_z \\ \dot{y}_n = -\frac{f^2+y_n^2}{f}\omega_x + \frac{x_n y_n}{f}\omega_y + x_n\omega_z \end{cases}, \tag{13}$$

and in the fine-tuning stage (third stage):

$$\begin{cases} \dot{x}_n = \frac{f}{Z}V_x - \frac{x_n}{Z}V_z - \frac{x_n y_n(t_0)}{f}\omega_x + \frac{f^2+x_n^2}{f}\omega_y - y_n\omega_z \\ \dot{y}_n = \frac{f}{Z}V_y - \frac{y_n}{Z}V_z - \frac{f^2+y_n^2}{f}\omega_x + \frac{x_n y_n}{f}\omega_y + x_n\omega_z \end{cases}. \tag{14}$$

To remove the noise in the image processing and feature extraction, a feature state estimator is designed based on the Kalman filter algorithm.

In the formulations below, k denotes the current time instant and $k+1$ the next time instant, while T_s represents the sampling time. The estimated states are denoted by ^ notation. Considering four features, the feature state at the current instant (k^{th} sample) is defined as:

$$X(k) = [x_1(k), y_1(k), ...x_4(k), y_4(k), \dot{x}_1(k), \dot{y}_1(k), ..., \dot{x}_4(k), \dot{y}_4(k)]^T, \tag{15}$$

or with consideration of (2):

$$X(k) = [s(k), \quad \dot{s}(k)], \tag{16}$$

where the elements of the vector can be obtained from (12), (13), or (14). Furthermore, the measurement vector represents the vector of the image feature points' coordinates extracted from the images of the camera:

$$M(k) = [x_{m1}(k), y_{m1}(k), ...x_{m4}(k), y_{m4}(k), \dot{x}_{m1}(k), \dot{y}_{m1}(k), ..., \dot{x}_{m4}(k), \dot{y}_{m4}(k)]^T \tag{17}$$

First, the prediction equations are:

$$\begin{aligned} \hat{X}(k|k-1) &= A\hat{X}(k-1|k-1) \\ P(k|k-1) &= AP(k-1|k-1)A^T + Q(k-1), \end{aligned} \tag{18}$$

where A is a 16×16 matrix whose diagonal elements equal one, $A_{i,i+8}(i = 1,2...,8)$ are equal to sampling time T_s, and the rest of the elements are zero, $P(k|k-1)$ represents the current prediction of the error covariance matrix, which gives a measure of the state estimate accuracy, while $P(k-1|k-1)$ is the previous error covariance matrix, and $Q(k-1)$ represents the process noise covariance computed using the information of the time instant $(k-1)$.

Second, the Kalman filter gain $D(K)$ is:

$$D(k) = P(k|k-1)(P(k|k-1) + R(k-1)^{-1}, \tag{19}$$

where $R(k-1)$ is the previous measurement covariance matrix.

Third, the estimation update is given as follows:

$$\begin{aligned} \hat{X}(k|k) &= \hat{X}(k|k-1) + D(k)(M(k) - \hat{X}(k|k-1) \\ P(k|k) &= P(K|k-1) - D(k)P(k|k-1), \end{aligned} \tag{20}$$

When the features are out of the FOV of the camera (i.e. $x_{mj}(k) = 0, y_{mj}(k) = 0, j = 1,2...4$), the feature reconstruction algorithm is proposed to provide the updated estimation vector under this circumstance. Since the features are out of FOV, the measurement vector will have some elements with zero values. This measurement vector will not lead to a satisfactory performance of switch IBVS. In order to improve the performance, instead of having zero values of the elements of $M(k)$ in (17), it is reasonable to assume that the nth feature that goes outside of FOV keeps its velocity at the moment (t_0) of leaving $(\dot{s}_n(t_0))$ during the period of feature loss. Hence, its position (i.e., point coordinates $s_n(t_0) = [x_{mn}(t_0), y_{mn}(t_0)]$) can be generated by integrating the velocity over the time. This means that the elements of $M(k)$ can be represented by this formulation:

$$M(k) = [(K_{ad} \sum_{l=0}^{b} \dot{s}_n(t_0)T_s + s_n(t_0)), \qquad \dot{s}_n(t_0)], \tag{21}$$

where $(l = 0,1,2,..,b)$ represents the number of time samples during the feature loss period, T_s is the sampling period, and k_{ad} is an adjusting coefficient. Once the feature is visible to the camera again, the actual value of $M(k)$ provided by the camera is used to replace the state estimation (21).

4. Controller Design

The IBVS controller was designed using the switch scheme. This method can set distinct gain values for the stages of the control law to achieve a fast response system while preserving the system stability.

In order to design the switch controller, the movement of the camera during the control task was divided into three different stages [15,16]. A criterion was needed for the switch condition between stages. In [15], the norm of feature errors was defined as the switching criterion. In this paper, a more intuitive and effective criterion is used [16]. As is shown in Figure 3, the switch angle criterion α is introduced as the angle between actual features and the desired ones. As soon as the angle α meets the predefined value, the controller law switches to the next stage.

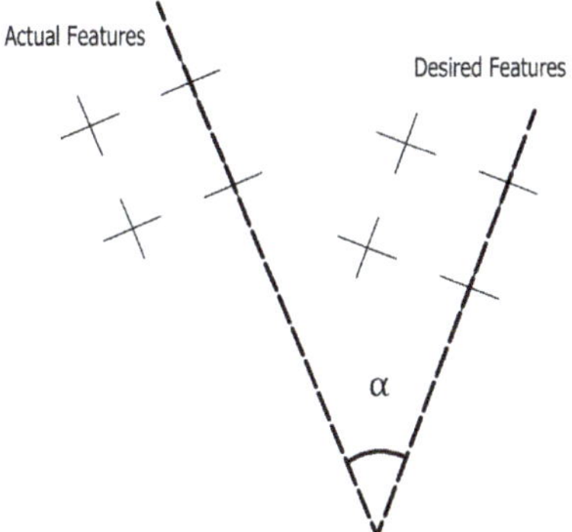

Figure 3. Switch angle criteria α: the angle between the desired and actual features.

Based on this criterion, the switching control law is presented as follows:

$$\begin{cases} V_{cs1} = -K_1 J_r^+ e(s), & |\alpha| \geq \alpha_0 \\ V_{cs2} = -K_2 J_t^+ e(s), & \alpha_1 \leq |\alpha| < \alpha_0 \ , \\ V_{cs3} = -K_3 J_{img}^+ e(s), & otherwise \end{cases} \tag{22}$$

where V_{csi} ($i = 1,2,3$) is the velocity of the camera in the ith stage, K_i is the symmetric positive definite gain matrix at each stage, and α_0 and α_1 are two predefined thresholds for the control law to switch to the next stage. The block diagram of the proposed algorithm is shown in Figure 4. Furthermore, the flowchart of the whole process of feature reconstruction and control is illustrated in Figure 5.

It was expected that in comparison with switch IBVS, the proposed method would ensure the smooth transition of the visual servoing task in the case of the feature loss and provide a better convergence performance.

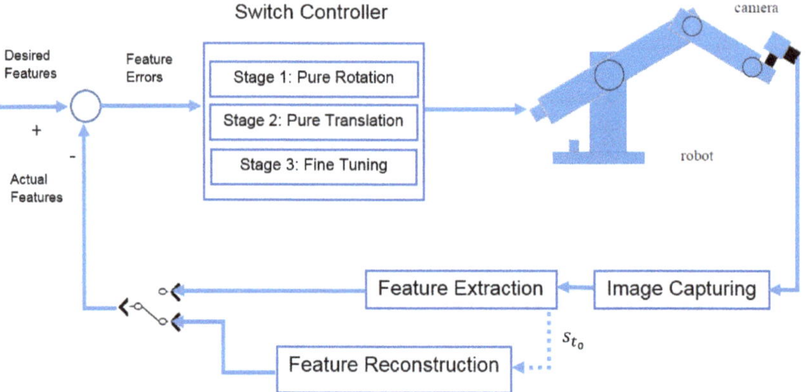

Figure 4. Block diagram of the proposed enhanced switch image-based visual servoing (IBVS) controller.

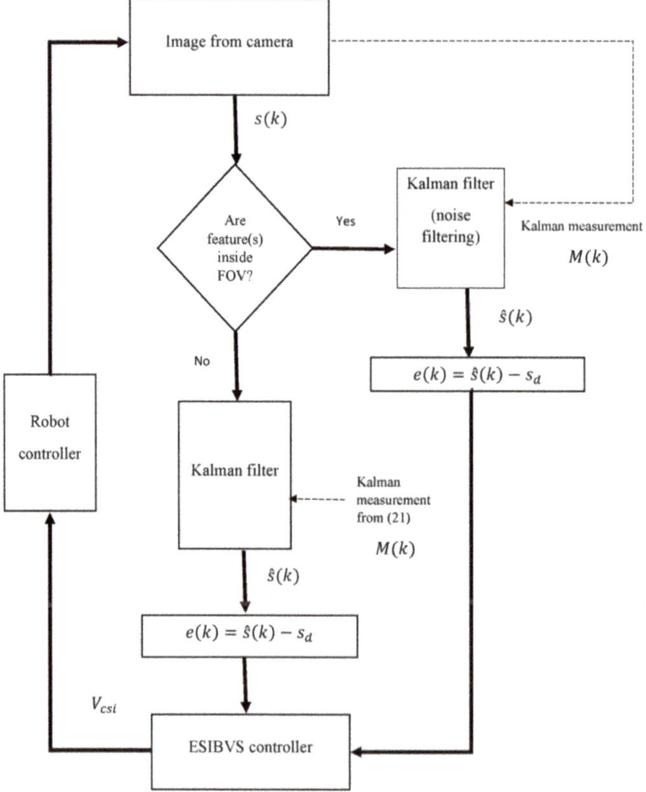

Figure 5. Flowchart of the Kalman filter feature reconstruction and control algorithm.

5. Simulation Results

To evaluate the performance of the proposed method, simulation tests were carried out by using MATLAB/SIMULINK software with the Vision and Robotic Toolbox. A 6-DOF DENSO robot with a camera installed in eye-in-hand configuration was simulated. The coordinates of the initial and desired features in the image space are given in Table 1. The camera parameters are as shown in Table 2.

Table 1. Test 1: simulation. Initial (I) and desired (D) feature point positions in pixels.

		Point 1		Point 2		Point 3		Point 4	
		(x, y)		(x, y)		(x, y)		(x, y)	
Test 1	I	376	757	202	621	20	814	218	969
	D	612	312	612	512	812	512	812	312

Table 2. Camera parameters.

Parameter	Value
Focal length (m) (f)	0.004
x axis scaling factor (pixel/m) (β)	110,000
y axis scaling factor (pixel/m) (β)	110,000
Principal point of x axis (pixel) (c_u)	120
Principal point of y axis (pixel) (c_v)	187

The task was to guide the end-effector to match the actual features with the desired ones in the camera image space. To simulate the condition where the features go outside FOV of the camera in real applications, the FOV of the camera was defined as the limited area shown in Figure 6a,b. When the features were in the defined FOV, they had actual position coordinates, and when they went outside FOV, the position coordinates of the features were set to zero. In this case, the proposed feature reconstruction algorithm was activated, and an estimate of the feature positions was generated. The norm of feature errors (NFE) is defined as below,

$$NFE = \sum_{n=1}^{4} \sqrt{(x_n - x_{dn})^2 + (y_n - y_{dn})^2}, \tag{23}$$

where x_n and y_n are the nth feature coordinates and x_{dn} and y_{dn} are the nth desired feature coordinates in the image plane.

In the simulation test, we set the initial feature coordinates and the desired ones in a way that the image features were out of FOV. Figures 6 and 7 demonstrate the performance comparison of the two methods. The paths of image features in the image space are given in Figure 6a,b. Figure 7a,b shows how the feature errors change with time in the proposed ESIBVSand switch method. Figure 7c,d demonstrates the norm of the feature errors' change with time in both methods. As shown in the figures, ESIBVS was able to reduce the norm of the errors to the preset threshold, while in the switch method, the norm of the errors did not converge. The summary of the simulation test is shown in Table 3. The results demonstrate how the proposed method was able to handle the situation in which the features went outside of the camera's FOV and completed the task successfully, while the switch method was unable to do so.

Table 3. Test 1: Comparison of simulation results between ESIBVSand switch IBVS.

	Time of Convergence (s)		Final Norm of Feature Errors (Pixel)	
	ESIBVS	Switch IBVS	ESIBVS	Switch IBVS
Test 1	12	Does not converge	1.5	Does not converge

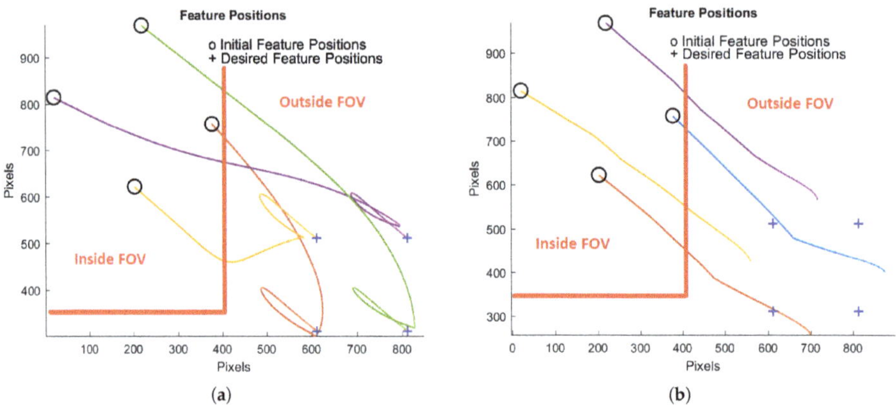

Figure 6. Test 1: simulation. Image space feature trajectory comparison of enhanced switch IBVS and switch IBVS. (**a**) Image space feature trajectory in enhanced switch IBVS; (**b**) image space feature trajectory in switch IBVS.

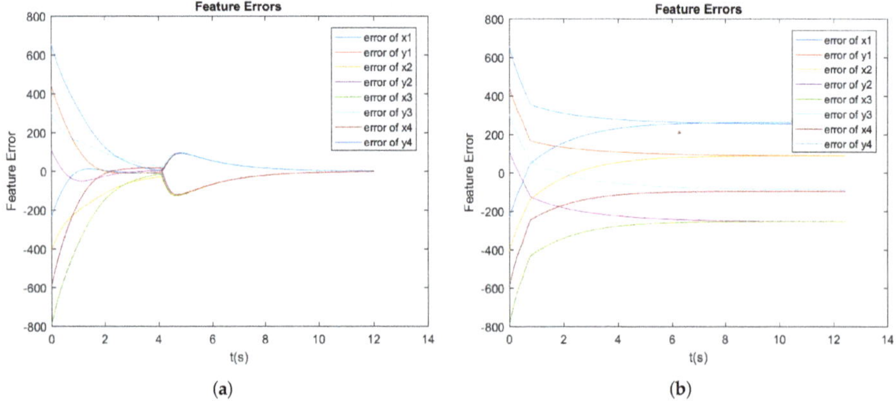

Figure 7. *Cont.*

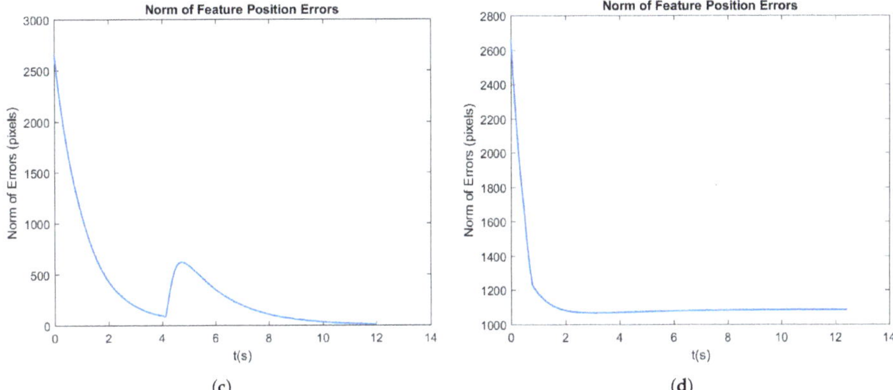

Figure 7. Test 1: simulation. Performance comparison of enhanced switch IBVS vs. switch IBVS. (**a**) Feature errors in enhanced switch IBVS; (**b**) feature errors in switch IBVS; (**c**) norm of feature errors in enhanced switch IBVS; (**d**) norm of feature errors in switch IBVS.

6. Experimental Results

In this section, to further verify the effectiveness of the proposed method, some experiments were carried out, and the results are presented. The experimental testbed included a 6-DOF DENSOrobot with a camera (specifications shown in Table 2) installed on its end-effector (Figure 8a). The camera model was a Logitech Webcam HD 720p, which captures the video with a resolution of 1280 × 720 pixels.

Two computers were used for the experimental tests. One computer carried out the image processing (PC2 in Figure 9) and sent the extracted feature coordinates to the other computer (PC1 in Figure 9), where the control algorithm was executed. Then, the control command (velocity of the end-effector) was sent to the robot controller. The image data taken by the camera were sent to an image processing program written by using the Computer Vision Toolbox of MATLAB. This program extracted the center coordinates of the features and sent them as feedback signals to the visual servoing controller in the sampling period of 0.001 s. Four feature points were used in the control task. The detailed information of the image processing and feature extraction algorithm can be seen in our previous work [32]. The goal was to control the end-effector so that the actual features matched the desired ones (Figure 8b).

To evaluate the efficiency of ESIBVS, its performance was compared to that of the switch IBVS method. In all the tests, the threshold value of NFE was set to 0.005 (equivalent to four pixels). When NFE reached this value, the robot stopped, and the servoing task was fulfilled. The initial angle α between the actual and desired features (Figure 3) was 50°. K_1, K_2, and K_3 in (22) were set to 1, 0.4, and 0.3, respectively.

Test 2: In this test (The video can be found in the Supplementary Materials), the initial and desired features were set such that they went outside of the FOV of the camera during the test. The initial and desired feature coordinates in the test are given in Table 4. Figure 10 demonstrates the movement of actual features during the test of ESIBVS. It illustrates how the features went outside of FOV, then were reconstructed, went back to FOV, and finally matched the desired features.

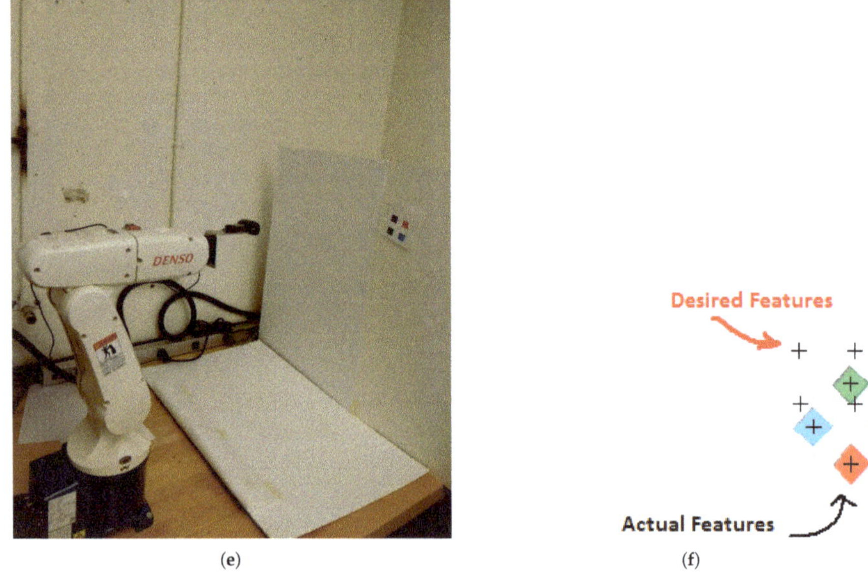

(e) (f)

Figure 8. (**a**) Experimental testbed, 6-DOF DENSO robot. (**b**) Actual and desired image features.

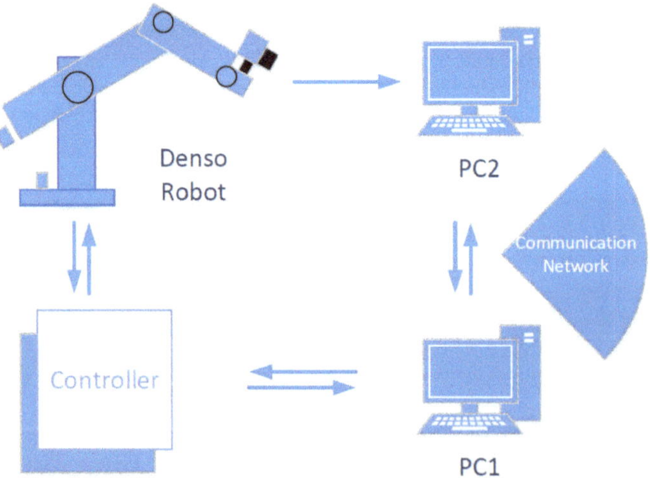

Figure 9. Structure of the experimental testbed.

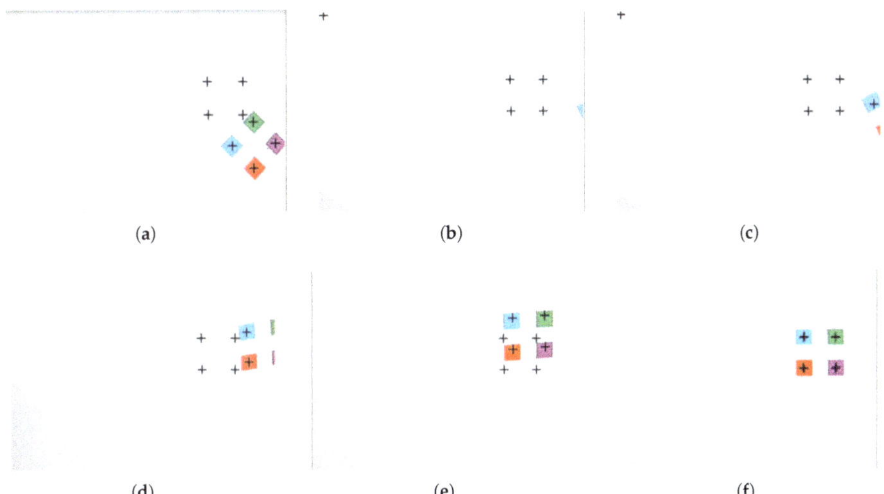

Figure 10. Test 2: Snap shots of the camera image during the enhanced switch IBVS test: (**a**) Desired and actual feature positions at the start. (**b**) Actual features are out of FOV. (**c**–**e**) Features are reconstructed and returned to FOV. (**f**) Final match of the desired and actual features.

Table 4. Test 2: experiment, Initial (I) and desired (D) feature point positions in pixels.

		Point 1		**Point 2**		**Point 3**		**Point 4**	
		(x, y)		**(x, y)**		**(x, y)**		**(x, y)**	
Test 2	I	251	132	278	102	306	127	279	157
	D	232	82	272	82	272	119	233	119

Figures 11–13 show the comparison results between ESIBVS and switch IBVS. Figure 11 shows the paths of features in the image space from the initial positions to the desired ones, as well as the camera trajectory in Cartesian space. In the proposed method, the actual and desired features matched, while in switch IBVS, the actual features did not converge to the desired ones. Figure 12 demonstrates the robot joint angles in ESIBVS and switch IBVS. Figure 13 shows the comparison regarding the feature errors. The feature errors and the norm of feature errors in the proposed method successfully converged to the desired values (Figure 13a,c), while in the switch IBVS, the task could not be completed, and thus, the feature errors did not converge (Figure 13b,d).

In order to further validate the performance of ESIBVS regarding the repeatability, the same test was repeated in 10 trials. The time of convergence and the final norms of feature error are shown in Table 5. The variations of feature error norms with time in 10 trials of ESIBVS are illustrated in Figure 14. As shown in the results, ESIBVS was able to overcome the feature loss and complete the task in each trial, while Switch IBVS was stuck in a point and did not converge.

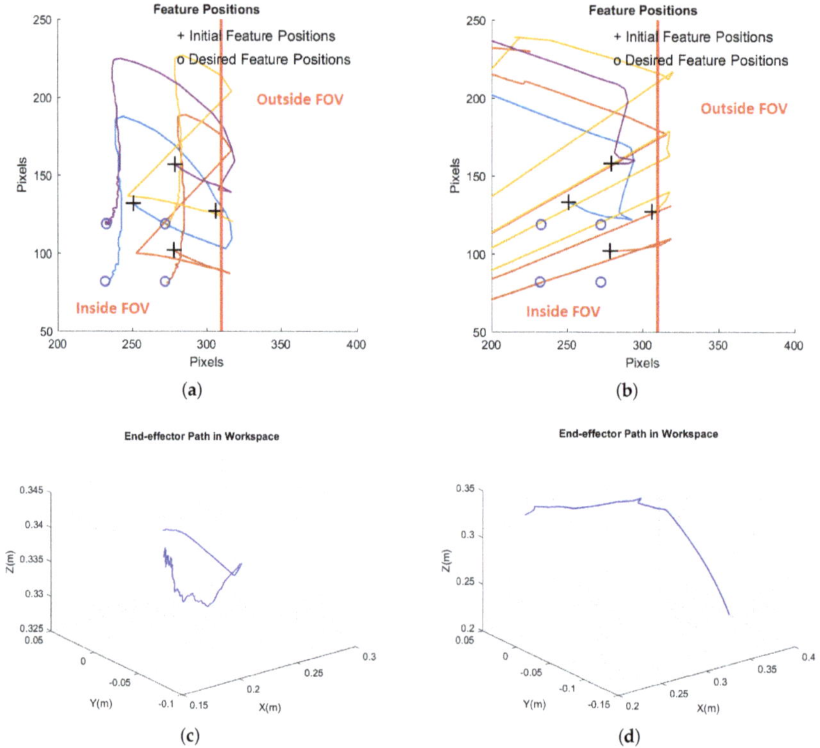

Figure 11. Test 2: experiment: Image space feature trajectory and 3D camera trajectory in enhanced switch IBVS and switch IBVS. (**a**) Image space feature trajectory in enhanced switch IBVS; (**b**) image space feature trajectory in switch IBVS; (**c**) camera 3D trajectory in enhanced switch IBVS; (**d**) camera 3D trajectory in switch IBVS.

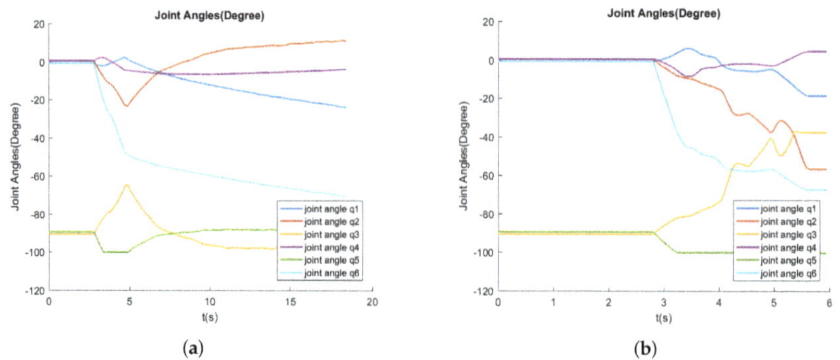

Figure 12. Test 2: experiment. Robot joint angles in enhanced switch IBVS and switch IBVS. (**a**) Joint angles (degree) in enhanced switch IBVS; (**b**) joint angles (degree) in switch IBVS.

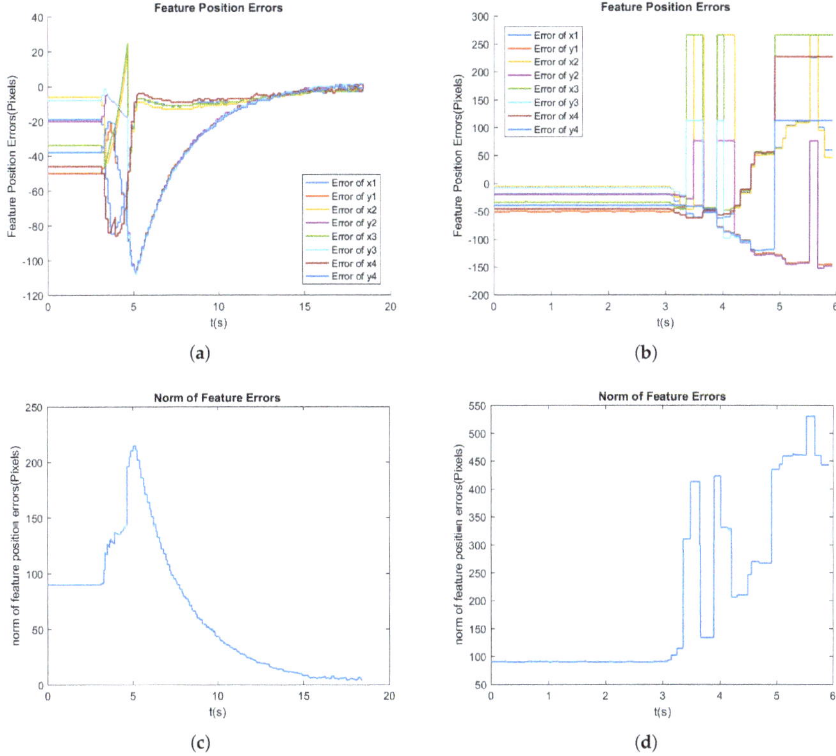

Figure 13. Test 2: experiment. Comparison of the feature errors and the norm of feature errors in enhanced switch IBVS and switch IBVS. (**a**) Feature errors in enhanced switch IBVS; (**b**) feature errors in switch IBVS; (**c**) norm of feature errors in enhanced switch IBVS; (**d**) norm of feature errors in switch IBVS.

Table 5. Test 2: experiment. Repeatability comparison results.

	Time of Convergence (s)		**Final Norm of Feature Errors (Pixel)**	
	ESIBVS	**Switch IBVS**	**ESIBVS**	**Switch IBVS**
Trial 1	19.95	Does not converge	3.4	Does not converge
Trial 2	18.99	Does not converge	3.1	Does not converge
Trial 3	17.75	Does not converge	2.8	Does not converge
Trial 4	17.94	Does not converge	3.7	Does not converge
Trial 5	19.37	Does not converge	3.6	Does not converge
Trial 6	18.29	Does not converge	2	Does not converge
Trial 7	20.34	Does not converge	3.1	Does not converge
Trial 8	19.03	Does not converge	3.4	Does not converge
Trial 9	18.77	Does not converge	2.4	Does not converge
Trial 10	18.74	Does not converge	3.4	Does not converge

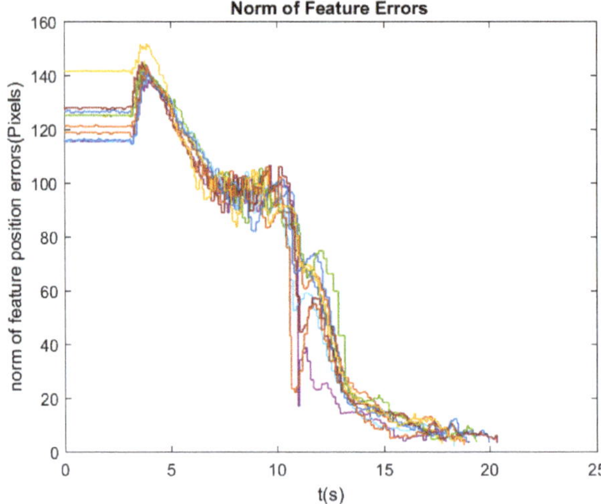

Figure 14. Test 2: experiment. The time variations of feature error norms in 10 trials of ESIBVS.

Test 3: In this test, the performance of ESIBVS was compared with that of switch IBVS in the situation where the features did not leave the FOV of the camera. The initial and desired features were set in a way such that the features did not go outside of FOV (Table 6). Similar to the previous tests, ESIBVS and switch IBVS were compared, and the results are shown in Figures 15–17 and Table 7. As shown in the figures, ESIBVS had a 38% shorter convergence time than switch IBVS did, which was owed to the superior noise-filtering ability of the designed Kalman filter.

Table 6. Test 3: Experiment. Initial (I) and desired (D) feature point positions in pixels.

		Point 1		Point 2		Point 3		Point 4	
		(x, y)		(x, y)		(x, y)		(x, y)	
Test 3	I	108	127	130	97	136	148	158	118
	D	232	82	272	82	272	119	233	119

Table 7. Test 3: Comparison of experimental resuts between ESIBVS and Switch IBVS.

	Time of Convergence (s)		Final Norm of Feature Errors (Pixel)	
	ESIBVS	Switch IBVS	ESIBVS	Switch IBVS
Test 3	8	12.5	3.4	3.6

The experimental results showed the efficiency of ESIBVS in dealing with feature loss while keeping the superior performance of the switch IBVS over traditional IBVS. As already shown in our previous work [15,16], the switch method was proven to have a better performance in its response time and its tracking performance, making it more feasible for industrial applications in comparison with the conventional IBVS. However, it suffered the drawback of weakness in dealing with feature loss. The proposed ESIBVS solved this problem and made switch IBVS more robust by using the Kalman filter to reconstruct the lost features.

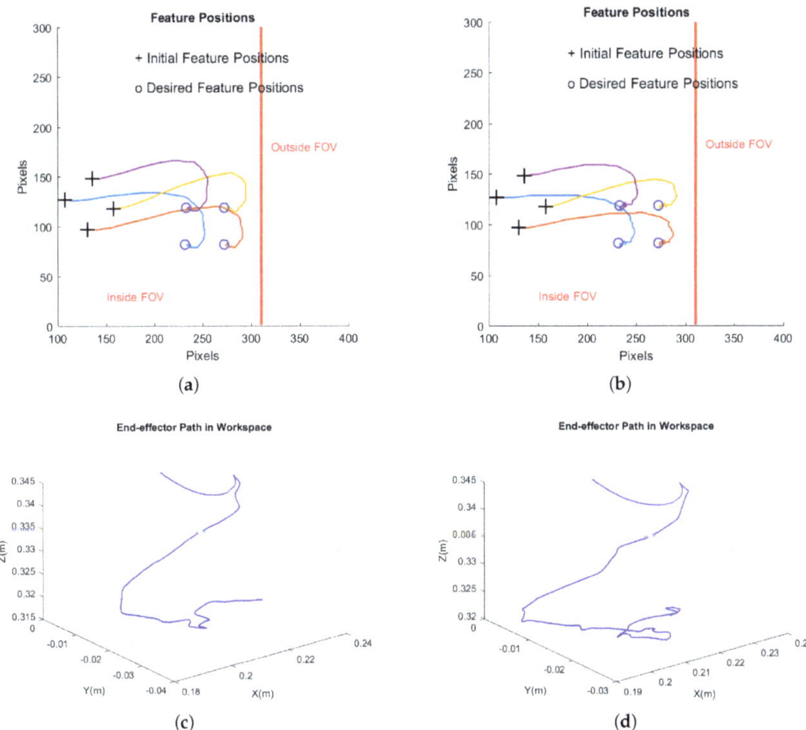

Figure 15. Test 3: experiment. Image space feature trajectory and 3D camera trajectory in enhanced switch IBVS and switch IBVS. (**a**) Image space feature trajectory in enhanced switch IBVS; (**b**) image space feature trajectory in switch IBVS; (**c**) camera 3D trajectory in enhanced switch IBVS; (**d**) camera 3D trajectory in switch IBVS.

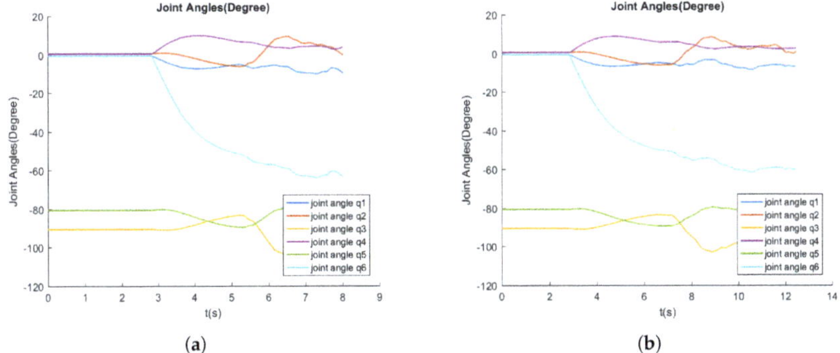

Figure 16. Test 3: experiment. Robot joint angles in enhanced switch IBVS and switch IBVS. (**a**) Feature errors in enhanced switch IBVS; (**b**) feature errors in switch IBVS.

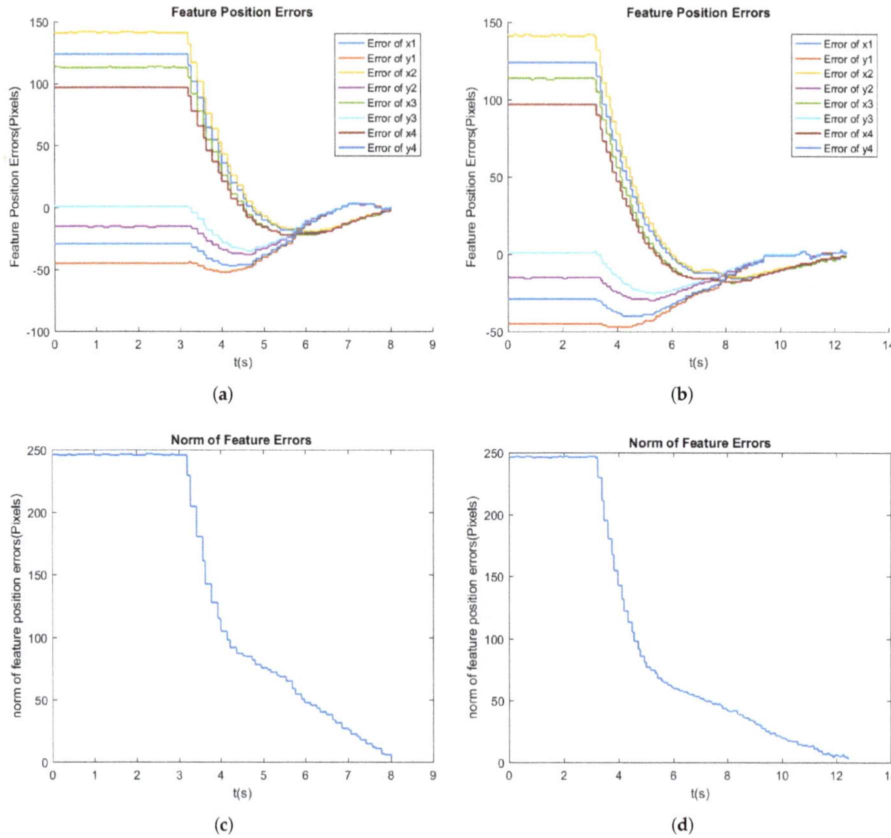

Figure 17. Test 3: experiment. Comparison of feature errors and the norm of feature errors in enhanced switch IBVS and switch IBVS. (**a**) Feature errors in enhanced switch IBVS; (**b**) feature errors in switch IBVS; (**c**) norm of feature errors in enhanced switch IBVS; (**d**) norm of feature errors in switch IBVS.

7. Conclusions

This paper proposed an enhanced switch IBVS for a 6-DOF industrial robot. An image feature reconstruction algorithm based on the Kalman filter was proposed to handle feature loss during the process of IBVS. The combination of a three-stage switch controller and feature reconstruction algorithm improved the system response speed and tracking performance of IBVS and simultaneously overcame the problem of feature loss during the task. The proposed method was simulated and then tested on a 6-DOF robotic system with the camera installed in an eye-in-hand configuration. Both simulation and experimental results verified the efficiency of the method. In the future, we may extend the method to make it more robust to uncertainties such as the depth of features and camera parameters. In addition, the effect of different sampling periods on the performance of the proposed ESIBVS will be investigated.

Supplementary Materials: The following are available online at http://www.mdpi.com/2079-9292/8/8/903/s1.

Author Contributions: The authors' individual contributions are provided as follow: conceptualization, A.G., P.L., W.-F.X. and W.T.; methodology, A.G., P.L., W.-F.X. and W.T.; software, A.G.; validation, A.G., P.L. and W.-F.X.; resources, W.-F.X.; writing–original draft preparation, A.G.; writing–review and editing, A.G., P.L., W.-F.X. and W.T.; supervision, W.-F.X. and W.T.; project administration, W.-F.X.; funding acquisition, W.-F.X.

Funding: This research was funded by Natural Sciences and Engineering Research Council of Canada grant number N00892.

Conflicts of Interest: The authors declare no conflict of interest. The funder had no role in the design of the study; in the collection, analyses, or interpretation of data; in the writing of the manuscript, or in the decision to publish the results.

References

1. Lee, Y.J.; Yim, B.D.; Song, J.B. Mobile robot localization based on effective combination of vision and range sensors. *Int. J. Control Autom. Syst.* **2009**, *7*, 97–104. [CrossRef]
2. Kim, J.H.; Lee, J.E.; Lee, J.H.; Park, G.T. Motion-based identification of multiple mobile robots using trajectory analysis in a well-configured environment with distributed vision sensors. *Int. J. Control Autom. Syst.* **2012**, *10*, 787–796. [CrossRef]
3. Banlue, T.; Sooraksa, P.; Noppanakeepong, S. A practical position-based visual servo design and implementation for automated fault insertion test. *Int. J. Control Autom. Syst.* **2014**, *12*, 1090–1101. [CrossRef]
4. Patil, M. Robot Manipulator Control Using PLC with Position Based and Image Based Algorithm. *Int. J. Swarm Intell. Evol. Comput.* **2017**, *6*, 1–8. [CrossRef]
5. Chaumette, F. Potential problems of stability and convergence in image-based and position-based visual servoing. In *The Confluence of Vision and Control*; Springer: London, UK, 1998; pp. 66–78.
6. Gans, N.R.; Hutchinson, S.A. Stable visual servoing through hybrid switched-system control. *IEEE Trans. Robot.* **2007**, *23*, 530–540. [CrossRef]
7. Malis, E.; Chaumette, F.; Boudet, S. 2 1/2 D visual servoing. *IEEE Trans. Robot. Autom.* **1999**, *15*, 238–250. [CrossRef]
8. Li, S.; Ghasemi, A.; Xie, W.F.; Gao, Y. An Enhanced IBVS Controller of a 6DOF Manipulator Using Hybrid PD-SMC Method. *Int. J. Control Autom. Syst.* **2018**, *16*, 844–855. [CrossRef]
9. Keshmiri, M.; Xie, W.F.; Ghasemi, A. Visual servoing using an optimized trajectory planning technique for a 4 DOFs robotic manipulator. *Int. J. Control Autom. Syst.* **2017**, *15*, 1362–1373. [CrossRef]
10. Keshmiri, M.; Xie, W.F. Image-based visual servoing using an optimized trajectory planning technique. *IEEE/ASME Trans. Mechatron.* **2016**, *22*, 359–370. [CrossRef]
11. Keshmiri, M.; Xie, W.F.; Mohebbi, A. Augmented image-based visual servoing of a manipulator using acceleration command. *IEEE Trans. Ind. Electron.* **2014**, *61*, 5444–5452. [CrossRef]
12. Kelly, R.; Carelli, R.; Nasisi, O.; Kuchen, B.; Reyes, F. Stable visual servoing of camera-in-hand robotic systems. *IEEE/ASME Trans. Mechatron.* **2000**, *5*, 39–48. [CrossRef]
13. Chen, J.; Dixon, W.E.; Dawson, M.; McIntyre, M. Homography-based visual servo tracking control of a wheeled mobile robot. *IEEE Trans. Robot.* **2006**, *22*, 406–415. [CrossRef]
14. Gans, N.R.; Hutchinson, S.A. A switching approach to visual servo control. In Proceedings of the 2002 IEEE International Symposium on Intelligent Control, Vancouver, BC, Canada, 30 October 2002; pp. 770–776.
15. Xie, W.F.; Li, Z.; Tu, X.W.; Perron, C. Switching control of image-based visual servoing with laser pointer in robotic manufacturing systems. *IEEE Trans. Ind. Electron.* **2009**, *56*, 520–529.
16. Ghasemi, A.; Xie, W.F. Decoupled image-based visual servoing for robotic manufacturing systems using gain scheduled switch control. In Proceedings of the 2017 International Conference on Advanced Mechatronic Systems (ICAMechS), Xiamen, China, 6–9 December 2017; pp. 94–99.
17. Hutchinson, S.; Hager, G.D.; Corke, P.I. A tutorial on visual servo control. *IEEE Trans. Robot. Autom.* **1996**, *12*, 651–670. [CrossRef]

18. Chaumette, F.; Hutchinson, S. Visual servo control. I. Basic approaches. *IEEE Robot. Autom. Mag.* **2006**, *13*, 82–90. [CrossRef]
19. Folio, D.; Cadenat, V. A controller to avoid both occlusions and obstacles during a vision-based navigation task in a cluttered environment. In Proceedings of the 44th IEEE Conference on Decision and Control, Seville, Spain, 15 December 2005; pp. 3898–3903.
20. Nicolis, D.; Palumbo, M.; Zanchettin, A.M.; Rocco, P. Occlusion-Free Visual Servoing for the Shared Autonomy Teleoperation of Dual-Arm Robots. *IEEE Robot. Autom. Lett.* **2018**, *3*, 796–803. [CrossRef]
21. Mezouar, Y.; Chaumette, F. Path planning for robust image-based control. *IEEE Trans. Robot. Autom.* **2002**, *18*, 534–549. [CrossRef]
22. Shen, T.; Chesi, G. Visual servoing path planning for cameras obeying the unified model. *Adv. Robot.* **2012**, *26*, 843–860.
23. Kazemi, M.; Gupta, K.K.; Mehrandezh, M. Randomized kinodynamic planning for robust visual servoing. *IEEE Trans. Robot.* **2013**, *29*, 1197–1211. [CrossRef]
24. Murao, T.; Yamada, T.; Fujita, M. Predictive visual feedback control with eye-in-hand system via stabilizing receding horizon approach. In Proceedings of the 2006 45th IEEE Conference on Decision and Control, San Diego, CA, USA, 13–15 December 2006; pp. 1758–1763.
25. Sauvée, M.; Poignet, P.; Dombre, E.; Courtial, E. Image based visual servoing through nonlinear model predictive control. In Proceedings of the 2006 45th IEEE Conference on Decision and Control, San Diego, CA, USA, 13–15 December 2006; pp. 1776–1781.
26. Lazar, C.; Burlacu, A.; Copot, C. Predictive control architecture for visual servoing of robot manipulators. In Proceedings of the IFAC World Congress, Milano, Italy, 28 August–2 September 2011; pp. 9464–9469.
27. Hajiloo, A.; Keshmiri, M.; Xie, W.F.; Wang, T.T. Robust online model predictive control for a constrained image-based visual servoing. *IEEE Trans. Ind. Electron.* **2016**, *63*, 2242–2250.
28. Assa, A.; Janabi-Sharifi, F. Robust model predictive control for visual servoing. In Proceedings of the 2014 IEEE/RSJ International Conference on Intelligent Robots and Systems (IROS 2014), Chicago, IL, USA, 14–18 September 2014; pp. 2715–2720.
29. Allibert, G.; Courtial, E.; Chaumette, F. Predictive control for constrained image-based visual servoing. *IEEE Trans. Robot.* **2010**, *26*, 933–939. [CrossRef]
30. García-Aracil, N.; Malis, E.; Aracil-Santonja, R.; Pérez-Vidal, C. Continuous visual servoing despite the changes of visibility in image features. *IEEE Trans. Robot.* **2005**, *21*, 1214–1220. [CrossRef]
31. Cazy, N.; Wieber, P.B.; Giordano, P.R.; Chaumette, F. Visual servoing when visual information is missing: Experimental comparison of visual feature prediction schemes. In Proceedings of the IEEE International Conference on Robotics and Automation (ICRA'15), Seattle, WA, USA, 26–30 May 2015; pp. 6031–6036.
32. Keshmiri, M. Image Based Visual Servoing Using Trajectory Planning and Augmented Visual Servoing Controller. Ph.D. Thesis, Concordia University, Montreal, QC, Canada, 2014.

Article

Visual Closed-Loop Dynamic Model Identification of Parallel Robots Based on Optical CMM Sensor

Pengcheng Li [1], Ahmad Ghasemi [1], Wenfang Xie [1,*] and Wei Tian [2]

[1] Department of Mechanical, Industrial & Aerospace Engineering, Concordia University, Montreal, QC H3G 1M8, Canada
[2] College of Mechanical and Electrical Engineering, Nanjing University of Aeronautics and Astronautics, Nanjing 210016, China
* Correspondence: wfxie@encs.concordia.ca

Received: 27 June 2019; Accepted: 22 July 2019; Published: 26 July 2019

Abstract: Parallel robots present outstanding advantages compared with their serial counterparts; they have both a higher force-to-weight ratio and better stiffness. However, the existence of closed-chain mechanism yields difficulties in designing control system for practical applications, due to its highly coupled dynamics. This paper focuses on the dynamic model identification of the 6-DOF parallel robots for advanced model-based visual servoing control design purposes. A visual closed-loop output-error identification method based on an optical coordinate-measuring-machine (CMM) sensor for parallel robots is proposed. The main advantage, compared with the conventional identification method, is that the joint torque measurement and the exact knowledge of the built-in robot controllers are not needed. The time-consuming forward kinematics calculation, which is employed in the conventional identification method of the parallel robot, can be avoided due to the adoption of optical CMM sensor for real time pose estimation. A case study on a 6-DOF RSS parallel robot is carried out in this paper. The dynamic model of the parallel robot is derived based on the virtual work principle, and the built dynamic model is verified through Matlab/SimMechanics. By using an outer loop visual servoing controller to stabilize both the parallel robot and the simulated model, a visual closed-loop output-error identification method is proposed and the model parameters are identified by using a nonlinear optimization technique. The effectiveness of the proposed identification algorithm is validated by experimental tests.

Keywords: parallel robot; dynamic model; visual servoing; closed-loop output-error identification; optical CMM sensor

---

1. Introduction

Parallel robots are closed-chain mechanisms in which the end-effector is supported by a series of independent computer-controlled serial chains linked to the base platform. Parallel robots present some outstanding advantages in higher force-to-weight ratio and better stiffness compared with serial manipulators. Specifically, 6-DOF parallel robots have been used in various applications (e.g., flight simulators [1], manufacturing lines [2] and medical tools [3]). Due to the manufacturing tolerances and deflection in the robot structure, the typical positioning discrepancy between a virtual robot in simulation and a real robot can be 8–15 mm [4], which does not meet the precision requirement of many potential applications. The low absolute accuracy of the robot is the main problem for the off-line programming based applications where tens of thousand points or continuous trajectories are to be reached or tracked.

The existence of closed-chain mechanism and multiple moving parts in the parallel robots, for example, in a 6-DOF Gough-Stewart platform consisting of 13 moving bodies (12 legs and one end-effector), yields difficulties in dynamic analysis of parallel robots. Moreover, the dynamic model plays an important role in the model-based controller designs, especially in applications where high positioning and tracking accuracy is needed. In contrast to serial robots, where the joint angles are used as the state variables for modeling, the parallel robots are preferably modeled and controlled in the operational (workspace) coordinates where the pose of the end-effector and its derivatives are used as the state variables [5]. The main reason is that there is almost no analytical expression for the forward kinematic model of 6-DOF parallel robots. For example, there are 40 forward kinematics solutions [6] mapping the joint coordinates to the pose of the end-effector platform in a Gough-Stewart platform. In addition, to this day, there is no effective method to determine the pose of the platform from multiple solutions. The dynamic models of 6-DOF Gough-Stewart parallel robot have been built by several approaches such as Newton-Euler [7], Lagrangian formulation [8] and the principle of virtual work [9]. Newton-Euler formulation can provide the internal forces for each individual body of the parallel robot, which can benefit the mechanical design process of the parallel robot. However, the computational load is high due to the large amount of equations. In contrast, the Lagrangian and virtual work methods are more efficient and suitable for the control design purpose since the reaction forces between the bodies of the parallel robot are not considered. The published research work on 6-DOF RSS parallel robots is considerably scarcer compared with that on Gough-Stewart parallel robot. The dynamic models of one type of 6-DOF RSS parallel robot, in which the active rotation axes are coplanar, are built based on Newton-Euler equations [10] or Lagrangian formulation [11] for dynamic analysis and tracking control purpose respectively. In this paper, the dynamic model of a 6-DOF RSS parallel robot, where the active rotation axes are parallel to each other, is built based on the virtual work principle, and the explicit form of the dynamic model is derived for identification and dynamic model-based visual servoing purposes.

The dynamic parameters are normally unknown or approximately derived from manufacturer specifications, which are not accurate enough for the dynamic model-based controller design. System identification is an effective method to perceive the uncertain parameters in the dynamic model of the system, and has been applied to many engineering practices [12]. As a highly coupled multi-input/multi-output (MIMO) nonlinear system, industrial robots aroused great interest and challenge for the identification method. The literatures on the state-of-the-art identification methods can be found in [13–15]. For the industrial robots, the dynamic identification is normally performed in closed-loop, since the robotic system is open loop unstable. In [16], a MIMO closed-loop identification based on weighted least square estimation has been applied to an industrial serial robot used in a planar configuration. In addition, other closed-loop identification methods with maximum likelihood, instrumental variable and related implementation issues on industrial serial robots are addressed in [17,18]. A new closed-loop output-error identification scheme has been adopted for the serial robots [19]. The output-error identification method aims at finding the dynamic model parameters by minimizing the output deviation between the actual and simulated systems subjecting to the same input [20]. In [19], the identification procedure is implemented in a closed-loop control structure and the joint torque is the measured output, which avoids the estimation of the velocity and acceleration from the measured joint position.

One potential issue of the above-mentioned identification methods is that joint torque measurement or a related control signal is needed for identification, which is not always available for the industrial robots, since the built-in controllers of many industrial robots are unaccessible and do not provide the torque actuation mode [21]. The input of built-in controllers is the position or velocity command, and the output is the joint torque which is unaccessible to the users. Hence the torque and current of the motors cannot be derived directly and it is not easy to install additional torque sensors to get the direct measurement. The unknown controller can be identified along with the dynamic parameters as introduced in [22]. However, joint torque measurement is still needed for identification purposes.

In [13,23–25], identification issues in parallel robots have been discussed. Most research works on dynamic identification of parallel robots are based on a simplified dynamic model, such as in [26–28]. Nevertheless, the systematic derivation of the full inverse dynamic model is proposed based on Jourdain's principle in [24]. In addition, the identification procedure is carried out in two steps: 1. identifying inertial parameters and 2. estimating the friction coefficients. In [23], the full parameters of robots' dynamic model and the joint drive gains are identified based on the total least square method, and the method is tested on a 3 DOF Orthoglide parallel robot. Many identification methods for 6 DOF parallel robots in [23–25] come directly from those of serial robots. By rewriting the inverse dynamic model and friction model into a linear form with respect to the dynamic and friction parameters, the identification of the unknown parameters can be done through least squares technique. However for 6 DOF parallel robots, to avoid solving forward kinematics, the regression matrix of the dynamic model is constituted of the pose state variables in the operational space. In [25], only joint space state variables are measured and estimated. Therefore the state variables in operational space should be calculated through numerical computation of the forward kinematics, which is quite time-consuming.

One way to avoid computing the forward kinematics in the identification process is to directly measure the pose of the end-effector in the operational space based on vision sensors. In [28], a camera-based dynamic identification procedure is given for a 4-DOF parallel robot with a heavily simplified dynamic model by considering only the inertia of the end-effector. Also, a vision sensor based dynamic identification has been carried out on serial robots [29] and cable robots [30], but seldom on parallel robots. The optical CMM sensor is a dual camera based vision sensor, which can provide the real-time pose information of the targets. The coordinates of target reflectors in the field of view (FOV) can be directly measured by the sensor. By observing four non-collinear reflectors on the platform, the pose of the end-effector can be measured. The optical CMM sensor has been applied to the kinematics calibration [31] and the path tracking controller design [32] for the robots. To increase the flexibility and tracking accuracy, vision can also be incorporated into the feedback control loop of the parallel robot systems to form the so-called visual servoing control. The pose of the end-effector can be acquired on-line by using the optical CMM sensor, i.e., C-track from Creaform Inc. (Levis, QC, Canada). The visual servoing controller for parallel robot is superior to the joint space controller due to the fact that the kinematic errors introduced from transforming the desired trajectory in the operational space into the one in the joint space in the joint controller can be avoided. The measured pose together with the visual servoing controller allows using the closed-loop output-error identification method [20] to identify the dynamic model of the parallel robot.

Upon the above discussions, a closed-loop output-error identification method based on a CMM sensor is proposed for parallel robots in this paper. The end-effector pose is measured by the optical CMM and served as the output of the real plant. The same outer loop visual servoing controller and reference trajectory are employed in both actual robot and simulation model for model identification. The forward kinematics of parallel robots, which is usually solved by using time-consuming numerical algorithm, can be avoided. The exact knowledge of the built-in controller and the joint torque are not needed. The dynamic model parameters are identified by using nonlinear optimization technique. The experimental tests validate the identification results.

This paper is organized as follows. Section 2 describes the dynamic model of a 6-RSS parallel robot. The closed-loop output-error identification method is proposed and the procedure of the identification is presented in Section 3. The dynamic model validation based on simulation and the experiment results of the identification are given in Section 4. Finally, the conclusion is drawn in Section 5.

2. Dynamic Modeling

In this section, the kinematic analysis of a 6 RSS parallel robot is conducted and the dynamic model is derived based on the virtual work principle.

2.1. Kinematic Analysis

The motivation of kinematic analysis is to determine geometry mapping from the motion of end-effector frame w.r.t the base frame (operational space motion) to the rotation of the actuators, as regarding the revolute joint frames (joint space motion). Based on the geometry mapping, the link Jacobian matrix is derived for building the dynamic model. As shown in Figure 1a, the 6-RSS parallel robot consists of six identical serial branch chains. Each serial branch, illustrated in Figure 1b, consists of a wrench, a link, an active revolute joint (R) and two passive spherical joints (S). One spherical joint is used to connect the wrench and the link. The revolute joint is driven by actuators and connects the wrench and the base platform, while the spherical joint is employed between the link and the end effector. The base frame ΣO is assigned at the symmetric center of the base platform and the end-effector frame ΣE is also attached at the symmetric center, while $\boldsymbol{E}$ denotes the coordinate vector of the frame origin regarding the base frame. The coordinate vectors of revolute joint centers regarding the base frame are marked by $\boldsymbol{B}_i (i = 1, 2, \ldots, 6)$ while the rotation centers of spherical joints are represented as $\boldsymbol{T}_i$ and $\boldsymbol{A}_i$ respectively. In the subsequent kinematic and dynamic analysis, the coordinates of the parts of parallel robot are defined regarding the base frame by default. The moving wrench and link frames ΣW_i and ΣL_i are attached to the wrenches and links respectively as depicted in Figure 1b. The coordinate vectors of the centers of mass of the wrenches and links are denoted as $\boldsymbol{c}_{w_i}$ and $\boldsymbol{c}_{l_i}$ respectively. The vector $\boldsymbol{\theta} = [\theta_1, \theta_2, \ldots, \theta_6]^T$ represents the rotation angles of the actuators. The coordinate vector from $\boldsymbol{E}$ pointing to $\boldsymbol{A}_i$ regarding the end-effector frame is denoted as $\boldsymbol{a}_i$. And the coordinate vectors $\boldsymbol{w}_i$ and $\boldsymbol{l}_i$ represent the directions and length of the wrench and link. The pose vector of the end-effector frame is expressed as $\boldsymbol{\chi}_E = [\boldsymbol{h}, \boldsymbol{\phi}]^T$, where $\boldsymbol{h} = [x, y, z]^T$ represents the position of the end-effector frame origin, while $\boldsymbol{\phi} = [\alpha, \beta, \gamma]^T$ represents the Euler-angle rotation of the frame. The rotation matrix, R, is given by

$$R = R_x(\alpha)R_y(\beta)R_z(\gamma) = \begin{pmatrix} c\beta c\gamma & -c\beta s\gamma & s\beta \\ c\alpha s\gamma + c\gamma s\beta s\alpha & -s\beta s\alpha s\gamma + c\alpha c\gamma & -c\beta s\alpha \\ s\alpha s\gamma - c\alpha c\gamma s\beta & c\alpha s\beta s\gamma + c\gamma s\alpha & c\beta c\alpha \end{pmatrix}, \tag{1}$$

where R_x, R_y, R_z are the orthonormal rotation matrices for the rotation about X, Y, Z axes respectively. The vector $\dot{\boldsymbol{\chi}}_E = \left[\dot{\boldsymbol{h}}^T, \dot{\boldsymbol{\phi}}^T\right]^T$ and $\ddot{\boldsymbol{\chi}}_E = \left[\ddot{\boldsymbol{h}}^T, \ddot{\boldsymbol{\phi}}^T\right]^T$ are the first and second order time derivatives of $\boldsymbol{\chi}_E$. The vector $\boldsymbol{v}_E = \left[\dot{\boldsymbol{h}}^T, \boldsymbol{\omega}^T\right]^T \in R^{6\times 1}$ denotes the linear and angular velocities of the end-effector frame. Then $\dot{\boldsymbol{v}}_E$ is the acceleration vector. The relationship between the Euler angle rate $\dot{\boldsymbol{\phi}}$ and angular velocity $\boldsymbol{\omega}$ is expressed as follows

$$\boldsymbol{\omega} = J_e\, \dot{\boldsymbol{\phi}}, \tag{2}$$

where $J_e = \begin{bmatrix} 1 & 0 & s\beta \\ 0 & c\alpha & -c\beta\, s\alpha \\ 0 & s\alpha & c\beta\, c\alpha \end{bmatrix}$ is the analytical Jacobian matrix, s and c stand for sin and cos respectively.

The following assumptions are made for kinematic and dynamic analysis of a 6-RSS parallel robot:

- The end-effector platform, wrenches and links are symmetric with respect to their axes.
- The links do not rotate about its symmetric axes.

As shown in Figure 1a,b, the closure loop position relationship between the end-effector frame and the base frame can be expressed as the following:

$$\boldsymbol{E} + \boldsymbol{a}_i - \boldsymbol{A}_i = 0 \quad and \tag{3}$$

$$\boldsymbol{B}_i + \boldsymbol{w}_i + \boldsymbol{l}_i - \boldsymbol{A}_i = 0, i = 1, 2, \ldots, 6. \tag{4}$$

Then the linear velocity can be derived by combining Equations (3) and (4) and taking the time derivative:

$$\dot{h} + \omega \times a_i = \omega_1 \times w_i + \omega_2 \times l_i, i = 1, 2, \ldots, 6, \tag{5}$$

where ω_1 and ω_2 are the angular velocity of frame ΣW_i and ΣL_i regarding ΣO respectively.

Then, dot multiplying l_i on both sides of Equation (5) yields:

$$l_i\dot{h} + (a_i \times l_i)\omega = (w_i \times l_i) \cdot \omega_1, i = 1, 2, \ldots, 6. \tag{6}$$

Since the wrench rotates around a fixed axis which is denoted as $\hat{s} = [0, 0, 1]^T$, the Jacobian matrix mapping from the end-effector Cartesian velocity to joint velocity can be derived by the following:

$$J_{a1}\dot{\theta} = J_{a2}v_E, \tag{7}$$

where $J_{a1} = diag((w_1 \times l_1) \cdot \hat{s}, (w_2 \times l_2) \cdot \hat{s}, \ldots, (w_6 \times l_6) \cdot \hat{s})$, $J_{a2} = \begin{bmatrix} l_1^T & (a_1 \times l_1)^T \\ \vdots & \vdots \\ l_6^T & (a_6 \times l_6)^T \end{bmatrix}$.

When the robot works in the singularity-free operational space, the Jacobian matrix J_{ad} can be derived as follows:

$$\dot{\theta} = J_{a1}^{-1} J_{a2} v_E = J_{ad} v_E. \tag{8}$$

Then the translational velocity of the center of mass of the wrench $\dot{c_{w_i}}$ can be obtained from Equation (9).

$$\dot{c_{w_i}} = \dot{\theta}_i \hat{s} \times c_{w_i} = J_{au} v_E, \tag{9}$$

where $J_{au} = (\hat{s} \times c_{w_i}) J_{adi} \in R^{3\times 6}$, and J_{adi} is the ith row of J_{ad}. Then the link Jacobian J_a mapping v_E to the velocity of the center of mass of the ith wrench $v_1 = [\dot{c_{w_i}}^T, \omega_{1i}^T]^T$ can be derived as

$$v_1 = \begin{bmatrix} \dot{c_{w_i}} \\ \omega_{1i} \end{bmatrix} = \begin{bmatrix} J_{au} \\ \hat{s} J_{adi} \end{bmatrix} v_E = J_a v_E. \tag{10}$$

Equation (11) can be deduced by right cross multiplying l_i on the both sides of Equation (5) and using Lagrange's rule.

$$(\omega_2 \cdot l_i) \cdot l_i - (l_i \cdot l_i) \cdot \omega_2 = \dot{h} \times l_i + (\omega \times a_i) \times l_i - (\omega_1 \times w_i) \times l_i. \tag{11}$$

Since the link does not rotate about its longitudinal axis, $\omega_2 \cdot l_i = 0$ holds. Rearranging Equation (11), the following equation can be derived:

$$\| l_i \|^2 \omega_2 = [l_i]_X \dot{h} - [l_i]_X [a_i]_X \omega + [l_i]_X [w_i]_X \omega_1, \tag{12}$$

where the operator $[\cdot]_X$ and $\| \cdot \|$ represents the cross product operation and Euclidean norm respectively.

By combining Equations (8) and (12), the following Jacobian matrix J_{bd} mapping from the end-effector Cartesian velocity to the angular velocity of the link frame can be deduced:

$$\begin{aligned} \omega_2 &= J_{bd} v_E, \\ J_{bd} &= \frac{1}{\| l_i \|^2} \{ \begin{bmatrix} [l_i]_X - [l_i]_X [a_i]_X \end{bmatrix} + [l_i]_X [w_i]_X \hat{s} J_{adi} \}. \end{aligned} \tag{13}$$

The translational velocity of the center of mass of the link $\dot{c_{l_i}}$ can be obtained as follows:

$$\dot{c_{l_i}} = -[w_i]_X \omega_1 - [c_{l_i}]_X \omega_2 \tag{14}$$

By substituting Equations (10) and (13) into Equation (14), the following velocity relationship can be derived:

$$\dot{c_{l_i}} = (-[w_i]_X \hat{s} J_{adi} - [c_{l_i}]_X J_{bd}) v_E = J_{bu} v_E \tag{15}$$

Hence the link Jacobian J_b mapping v_E to the velocity of the center of mass of the ith link v_2 can be derived.

$$v_2 = \begin{bmatrix} \dot{c_{l_i}} \\ \omega_{2i} \end{bmatrix} = \begin{bmatrix} J_{bu} \\ J_{bd} \end{bmatrix} v_E = J_b v_E \tag{16}$$

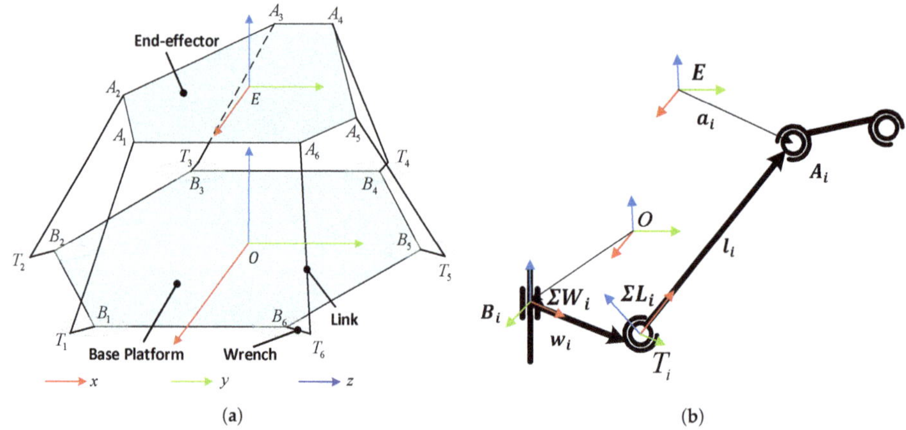

Figure 1. (**a**) The sketch of 6-RSS parallel robot, (**b**) Single serial branch.

2.2. Dynamic Modeling

In contrast to serial robots, the dynamic modeling of parallel robots is more complicated due to its closure geometrical structure and difficulties in deriving the forward kinematics. The principle of virtual work is employed to derive the explicit form of the dynamic model in terms of the operational coordinates and their time derivatives as shown in Equation (17) for 6-DOF RSS robot, which is useful for dynamic model-based controller design.

$$\tau_g = M(\chi_E)\dot{v}_E + C(\chi_E, v_E)v_E + G(\chi_E) + \tau_f, \tag{17}$$

where τ_g denotes the general force acting on the end-effector frame, τ_f is the friction, $M(\chi_E)$ is the mass matrix, $C(\chi_E, v_E)$ is Coriolis and centrifugal matrix, and $G(\chi_E)$ denotes the gravity. In order to avoid solving the forward kinematics of the parallel robot which may not have analytical solutions, the pose in the coordinates of the end-effector and its time derivatives are employed in Equation (17).

The balance equation of virtual work principle for a moving rigid body, $*$, can be expressed as follows:

$$\bar{F}_* \cdot \delta\chi_* = (F_{ext_*} + \tilde{F}_*) \cdot \delta\chi_* = 0, \tag{18}$$

in which $\bar{F}_*$ contains the static balance force and torque, $F_{ext_*} = [f_{ext_*}^T, \tau_{ext_*}^T]^T$ is the external force (f_{ext_*}) and torque (τ_{ext_*}) exerted on the center of mass of the body respectively, $\delta\chi_*$ denotes the virtual displacement, and the fictitious force and torque are:

$$\tilde{F}_* = \begin{bmatrix} m_* g - m_* \ddot{h}_* \\ -(I_* \dot{\omega}_* + \omega_* \times I_* \omega_*) \end{bmatrix}, \tag{19}$$

where m_* is the mass of the body, $\boldsymbol{g}$ is the gravity vector, $\ddot{\boldsymbol{h}}_*$ is the linear acceleration of center of mass of the body, I_* is the moment of inertia, $\boldsymbol{\omega}_*$ and $\dot{\boldsymbol{\omega}}_*$ denote the angular velocity and acceleration of the moving body frame. Further, $\bar{\boldsymbol{F}}_*$ can be represented in the form similar to Equation (17):

$$
\begin{aligned}
&\bar{\boldsymbol{F}}_* = \boldsymbol{F}_{\boldsymbol{ext}_*} + M_*\dot{\boldsymbol{v}}_* + C_*\boldsymbol{v}_* + G_*, \ where \\
&M_* = \begin{bmatrix} -m_* E_{3\times3} & 0_{3\times3} \\ 0_{3\times3} & -I_* \end{bmatrix}, C_* = \begin{bmatrix} 0_{3\times3} & 0_{3\times3} \\ 0_{3\times3} & -\boldsymbol{\omega}_* \times I_* \end{bmatrix}, \\
&G_* = \begin{bmatrix} m_*\boldsymbol{g} \\ 0_{3\times1} \end{bmatrix}, \dot{\boldsymbol{v}}_* = \begin{bmatrix} \ddot{\boldsymbol{h}}_* \\ \dot{\boldsymbol{\omega}}_* \end{bmatrix}, \boldsymbol{v}_* = \begin{bmatrix} \dot{\boldsymbol{h}}_* \\ \boldsymbol{\omega}_* \end{bmatrix}
\end{aligned}
\tag{20}
$$

in which $E_{3\times3} \in R^{3\times3}$ denotes the identity matrix. For a 6-DOF RSS parallel robot, there are 13 moving bodies including the end-effector, 6 wrenches and 6 links. Therefore the balance equation of the 6-DOF parallel robot can be rewritten as Equation (21):

$$
\bar{\boldsymbol{F}}_p \cdot \delta\boldsymbol{\chi}_{\boldsymbol{e}} + \sum_{i=1}^{6} \bar{\boldsymbol{F}}_{l_i} \cdot \delta\boldsymbol{\chi}_{\boldsymbol{l}_i} + \sum_{i=1}^{6} \bar{\boldsymbol{F}}_{w_i} \cdot \delta\boldsymbol{\chi}_{\boldsymbol{w}_i} = 0, \tag{21}
$$

where $\bar{\boldsymbol{F}}_p$, $\bar{\boldsymbol{F}}_{l_i}$ and $\bar{\boldsymbol{F}}_{w_i}$ contain the static balance force and torque exerted on the centers of mass of the platform, links and wrenches respectively and can be represented in the same form as Equation (20), $\delta\boldsymbol{\chi}_{\boldsymbol{e}}$, $\delta\boldsymbol{\chi}_{\boldsymbol{l}_i}$, and $\delta\boldsymbol{\chi}_{\boldsymbol{w}_i}$ are the virtual displacements accordingly.

In addition, the following relations hold for the velocity analysis:

$$
\delta\boldsymbol{\theta} = J_{ad}\delta\boldsymbol{\chi}_{\boldsymbol{e}}, \ \ \delta\boldsymbol{\chi}_{\boldsymbol{w}_i} = J_a\delta\boldsymbol{\chi}_{\boldsymbol{e}}, \ \ \delta\boldsymbol{\chi}_{\boldsymbol{l}_i} = J_b\delta\boldsymbol{\chi}_{\boldsymbol{e}}. \tag{22}
$$

Substituting Equation (22) into Equation (21), the terms in Equation (17) can be derived as the following:

$$
\begin{aligned}
M(\boldsymbol{\chi}_{\boldsymbol{E}}) &= M_p + \sum_{i=1}^{6}(J_{a_i}^T M_{w_i} J_{a_i} + J_{b_i}^T M_{l_i} J_{b_i}), \\
C(\boldsymbol{\chi}_{\boldsymbol{E}}, \boldsymbol{v}_{\boldsymbol{E}}) &= C_p + \sum_{i=1}^{6}(J_{a_i}^T C_{w_i} J_{a_i} + J_{a_i}^T M_{w_i} \dot{J}_{a_i} + J_{b_i}^T C_{l_i} J_{b_i} + J_{b_i}^T M_{l_i} \dot{J}_{b_i}), \\
G(\boldsymbol{\chi}_{\boldsymbol{E}}) &= G_p + \sum_{i=1}^{6}(J_{a_i}^T G_{w_i} + J_{b_i}^T G_{l_i}), \\
\boldsymbol{\tau}_{\boldsymbol{g}} &= J_{ad}^T \boldsymbol{\tau}_{\boldsymbol{a}},
\end{aligned}
\tag{23}
$$

where $\boldsymbol{\tau}_{\boldsymbol{a}} = [\tau_{a_1}, \tau_{a_2} ... \tau_{a_6}]^T$ is the actuator torque vector applying on the revolute joints, and the details of Equation (23) are given in Appendix A. The joint friction is described by Coulomb model [33] that has been used in the modelings of both parallel robots [24] and serial robots. Based on this friction model, the friction $\boldsymbol{\tau}_{\boldsymbol{f}}$ in Equation (17) can be represented as:

$$
\boldsymbol{\tau}_{\boldsymbol{f}} = J_{ad}^T \begin{bmatrix} f_{c_1} sign(J_{ad_1}\boldsymbol{v}_{\boldsymbol{E}}) + f_{v_1} J_{ad_1}\boldsymbol{v}_{\boldsymbol{E}} \\ f_{c_2} sign(J_{ad_2}\boldsymbol{v}_{\boldsymbol{E}}) + f_{v_2} J_{ad_2}\boldsymbol{v}_{\boldsymbol{E}} \\ \cdots \\ f_{c_6} sign(J_{ad_6}\boldsymbol{v}_{\boldsymbol{E}}) + f_{v_6} J_{ad_6}\boldsymbol{v}_{\boldsymbol{E}} \end{bmatrix}, \tag{24}
$$

where f_{c_i} and f_{v_i} are the Coulomb and viscous friction parameters of the ith revolute joint.

2.3. Dynamic Model Simplification

Since the kinematic parameters of parallel robots can be obtained from the manufacturer specifications or by calibration [31], only inertia and friction parameters are considered for the model identification of parallel robots. Considering the heavy computation load of solving inverse dynamic model for the dynamic identification and visual servoing purpose, a reduction of the dynamic parameters with the trade-off between the computation load and accuracy should be implemented. The geometry feature of the parallel robot can be considered for simplification. For the 6-RSS parallel robot, it is assumed that the wrenches, links and end-effector are symmetric. Furthermore, the center of mass is assumed to be located in the symmetric center. Therefore, the dynamic parameters for each body of the wrenches, links and end-effector can be reduced to $\boldsymbol{\xi}_* = [m_*, I_{x_*}, I_{y_*}, I_{z_*}]$. Then Equation (17) can be rewritten as the linear form w.r.t. the dynamic parameters and the friction coefficients:

$$\tau_g = \Gamma(\chi_E, v_E, \dot{v}_E)\Xi, \tag{25}$$

where $\Xi = [\xi_p^T, \xi_{w1}^T, \xi_{w2}^T, \ldots, \xi_{w6}^T, \xi_{l1}^T, \xi_{l2}^T, \ldots, \xi_{l6}^T, f_{c_1}, f_{c_2}, \ldots, f_{c_6}, f_{v_1}, f_{v_2}, \ldots, f_{v_6}]^T$ is a $R^{64\times 1}$ vector of dynamic parameters and the friction coefficients, and $\Gamma(\chi_E, v_E, \dot{v}_E)$ is the regressor matrix, which consists of the kinematic parameters, state variables and their derivatives. $\Gamma(\chi_E, v_E, \dot{v}_E)$ can be derived using the Symbolic Math Toolbox of Matlab. Given an exciting trajectory as a reference input to the robot, which will be introduced in Section 3.3, Equation (26) can be obtained by reorganizing Equation (25).

$$\begin{bmatrix} \tau_{g_1} \\ \tau_{g_2} \\ \vdots \\ \tau_{g_n} \end{bmatrix} = \begin{bmatrix} \Gamma(\chi_{E_1}, v_{E_1}, \dot{v}_{E_1}) \\ \Gamma(\chi_{E_2}, v_{E_2}, \dot{v}_{E_2}) \\ \vdots \\ \Gamma(\chi_{E_n}, v_{E_n}, \dot{v}_{E_n}) \end{bmatrix} \Xi = H\Xi, \tag{26}$$

where n is the number of the sampled poses from the given trajectory. By feeding various testing trajectories to the robot, the regression matrix H is of full rank which means all elements of Ξ can be identified.

3. Closed-Loop Output-Error Identification Based on Vision Feedback

3.1. Pose Estimation Using Optical CMM

As shown in Figure 2, a dual-camera optical CMM C-track 780 is employed to measure the pose of end-effector for the identification of the dynamic model in this research. The pose measurement principle of the optical CMM sensor is presented in this subsection. The target reflectors are taken as the point features. The homogeneous coordinates of the reflectors w.r.t. the sensor frame can be obtained by using triangulation principle [34]. Given a group of non-collinear reflectors $p_i (i = 1, 2, \ldots, n)$ attached on the end-effector platform, the homogeneous coordinates in the sensor frame can be measured and denoted as ${}^C\mathbf{P}_i = [x_{pi}, y_{pi}, z_{pi}, 1]^T$. In addition, the homogeneous coordinates of the reflectors w.r.t. the end-effector frame ΣE denoted as ${}^E\mathbf{P}_i = ({}^E x_i, {}^E y_i, {}^E z_i, 1)$ are known from the definition of the end-effector frame ΣE. Correspondingly, the transformation equation of ith feature point can be written as:

$${}^C\mathbf{P}_i = {}^C\mathbf{T}_E\, {}^E\mathbf{P}_i, \tag{27}$$

where ${}^C\mathbf{T}_E$ is the homogeneous transformation matrix mapping from ΣE to ΣC. In order to derive ${}^C\mathbf{T}_E$, at least three non-collinear feature points are required [35]. However, as indicated in [36], at least four coplanar feature points are needed for obtaining a unique solution while additional non-coplanar feature points can be used to improve the estimation accuracy with measurement noise. Similarly, the homogeneous transformation matrix mapping from ΣO to ΣC, ${}^C\mathbf{T}_O$, can be derived. Then the pose of the end-effector frame w.r.t. the base frame, χ_E, can be obtained from the homogeneous transformation matrix ${}^{O}\mathbf{T}_E$ given by Equation (28).

$$ {}^{O}\mathbf{T}_E = {}^{C}\mathbf{T}_O^{-1}\,{}^{C}\mathbf{T}_E. \tag{28} $$

By using the proprietary software VXelements provided by Creaform Inc., the target frames can be defined based on the selected reflectors on the surface of the end-effector and base frame respectively. The real-time position and rotation information of the end-effector frame w.r.t. the base frame can be acquired, recorded or displayed simultaneously. In addition, the computation associated with the pose estimation of the target frame is carried out by VXelements.

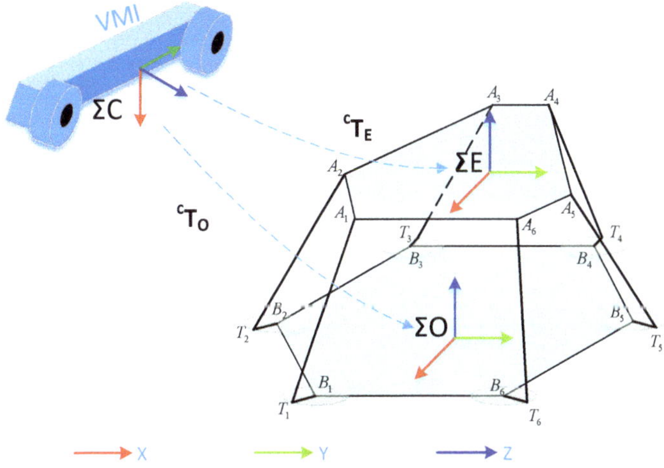

Figure 2. The pose measurement system of a 6-RSS parallel robot.

3.2. Closed-Loop Output-Error Identification Method

The basic idea of output error identification is to use nonlinear optimization technique to minimize the squared error between the output of the real plant and the simulated model. Since the motor torque is unaccessible, the pose of the end-effector, χ_E, measured by optical CMM is used as the output of the system. Hence the closed-loop output-error identification method is adopted [19]. In conventional closed-loop output-error identification method, the controllers should be exactly known and applied to both the real plant and the simulated model. However for an industrial robot, the built-in controllers are usually unknown and need to be identified.

The closed-loop output-error identification approach for vision based robotic system, as depicted in Figure 3, is proposed in this paper. For the built-in controller of the 6-RSS parallel robot, a PID controller is used to control the joint angle of each revolute joint. However the three gains of PID controller are unknown and needed to be identified. During the process of identification, the gains and dynamic parameters are updated in each iteration of nonlinear optimization, which may make the simulated model unstable. Hence, an outer loop visual servoing controller is added to stabilize both real robot and simulated model. The visual servoing controller and the built-in controller construct a cascade PID controller. As stated in [37,38], the cascade controller yields better dynamic performance in terms of stability and working frequency compared with single loop controller. With a well-tuned outer loop visual servoing controller, both the real and simulation systems can have a better performance and stability.

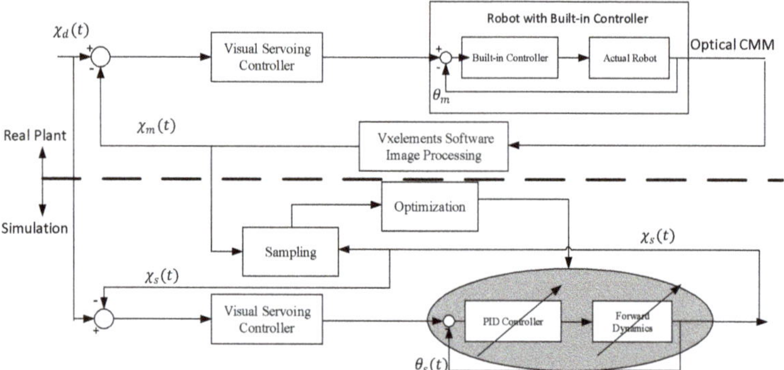

Figure 3. The block diagram of closed-loop control system for model identification of 6-RSS parallel robot.

In Figure 3, for both the real plant and simulated model, the same exciting trajectory is given as the input $\chi_d(t)$. Then the outer loop visual servoing controller is designed as Equation (30). The inverse kinematic Jacobian J_θ is used to transform the velocity in operational space to that in joint space. By combining Equations (2) and (8), Jacobian matrix J_θ is derived as shown in Equation (29) and the joint position control signal $\boldsymbol{U}_\theta(t)$ is generated as shown in Equation (30).

$$\dot{\boldsymbol{\theta}} = J_{ad}\begin{bmatrix} E_{3\times 3} & \mathbf{0}_{3\times 3} \\ \mathbf{0}_{3\times 3} & J_e \end{bmatrix}\dot{\chi}_E = J_\theta \dot{\chi}_E. \tag{29}$$

$$\boldsymbol{U}_\theta(t) = J_\theta[k_p(\chi_d(t) - \chi_m(t)) + k_d(\dot{\chi}_d(t) - \dot{\chi}_m(t)) + k_i\int(\chi_d(t) - \chi_m(t))dt], \tag{30}$$

where k_i, k_p, k_d are constants, and $\chi_m(t)$ is the visual feedback obtained from VXelements software, while in the simulation $\chi_m(t)$ is replaced by $\chi_s(t)$ which is the pose calculated by using the forward dynamic model.

In addition, then the PID controller, given in Equation (31), is used to describe the built-in joint controller in each joint and is employed in the simulated model.

$$\boldsymbol{\tau}_a(t) = l_p(\boldsymbol{U}_\theta(t) - \boldsymbol{\theta}_s(t)) + l_d(\dot{\boldsymbol{U}}_\theta(t) - \dot{\boldsymbol{\theta}}_s(t)) + l_i\int(\boldsymbol{U}_\theta(t) - \boldsymbol{\theta}_s(t))dt, \tag{31}$$

where l_i, l_p, l_d are the PID parameters to be identified, and $\boldsymbol{\theta}_s(t)$ is the joint positions, which can be obtained by analytically solving the inverse kinematics.

The real plant output $\boldsymbol{Y}_m = [\chi_m(1), \chi_m(2), \cdots, \chi_m(k),]^T$ and the simulation output $\boldsymbol{Y}_s = [\chi_s(1), \chi_s(2), \cdots, \chi_s(k)]^T$ are fed to the optimization process. The parameters to be identified, $\boldsymbol{\Lambda}$, can be denoted as $\boldsymbol{\Lambda} = [\Xi^T, l_p, l_i, l_d]^T$. Accordingly the identification of $\boldsymbol{\Lambda}$ can be converted to solving the following nonlinear optimization problem:

$$Minimize\ \Phi(\boldsymbol{\Lambda}) = \parallel \boldsymbol{Y}_m - \boldsymbol{Y}_s \parallel^2. \tag{32}$$

Then the updating formula for $\boldsymbol{\Lambda}$ is given as follows:

$$\boldsymbol{\Lambda}^{r+1} = (J_\Phi^T J_\Phi)^{-1} J_\Phi^T \Phi(\boldsymbol{\Lambda}^r) + \boldsymbol{\Lambda}^r \tag{33}$$

where $\boldsymbol{\Lambda}^r$ is the value of $\boldsymbol{\Lambda}$ in the rth iteration and J_Φ is the Jacobian matrix of $\Phi(\boldsymbol{\Lambda})$ w.r.t. $\boldsymbol{\Lambda}$ given as:

$$J_\Phi = \begin{bmatrix} \frac{\partial\Phi(\boldsymbol{\Lambda})}{\partial\boldsymbol{\Lambda}_1} & \frac{\partial\Phi(\boldsymbol{\Lambda})}{\partial\boldsymbol{\Lambda}_2} & \cdots & \frac{\partial\Phi(\boldsymbol{\Lambda})}{\partial\boldsymbol{\Lambda}_{67}} \end{bmatrix}, \tag{34}$$

where $\mathbf{\Lambda}_i$ denotes the ith column of $\mathbf{\Lambda}$. The terminate criteria is given as:

$$\begin{aligned} &\frac{\| \Phi(\mathbf{\Lambda}^{r+1}) - \Phi(\mathbf{\Lambda}^{r}) \|}{\| \Phi(\mathbf{\Lambda}^{r}) \|} \leq tol_1 \\ &\max_{i=1,\cdots,n} \left| \frac{\mathbf{\Lambda}_i^{r+1} - \mathbf{\Lambda}_i^{r}}{\mathbf{\Lambda}_i^{r}} \right| \leq tol_2, \end{aligned} \tag{35}$$

where $|\cdot|$ denotes the absolute-value norm operation, tol_1 and tol_2 are the thresholds to be chosen for tuning the accuracy. A compromise should be made between the convergence speed and accuracy when choosing thresholds. To achieve good results in solving the nonlinear optimization problem, a proper initial guess of $\mathbf{\Lambda}$ is needed. For the dynamic parameters of the parallel robots, the initial guess can be calculated from manufacturer specifications. Then a priori PID parameters are obtained based on the simulation model.

3.3. Modified Exciting Trajectory

The exciting trajectories for the dynamic model identification of the serial robots based on the inverse dynamic model have been well studied in [39]. The Finite Fourier series-based exciting trajectory has been tested in a large amount of research works for identification purpose [40]. For serial robots, it can be represented by the following:

$$\begin{aligned} \theta_i(t) &= \sum_{l=1}^{n} [\frac{sin(2\pi f_0 lt)}{2\pi f_0 l} s_i^l - \frac{cos(2\pi f_0 lt)}{2\pi f_0 l} c_i^l] + \theta_{0_i} \\ \dot{\theta}_i(t) &= \sum_{l=1}^{n} [cos(2\pi f_0 lt) s_i^l + sin(2\pi f_0 lt) c_i^l] \\ \ddot{\theta}_i(t) &= \sum_{l=1}^{n} [-2\pi f_0 l sin(2\pi f_0 lt) s_i^l + 2\pi f_0 lt cos(2\pi f_0 lt) c_i^l], \end{aligned} \tag{36}$$

where $\theta_i(t)$ is the ith joint angle trajectory of serial robots, n is the harmonics number, f_0 is the fundamental frequency, and s_i^l, c_i^l, θ_{0_i} are the trajectory parameters to be optimized. Instead of choosing the joint space states $(\boldsymbol{\theta}, \dot{\boldsymbol{\theta}}, \ddot{\boldsymbol{\theta}})$ for serial robots, the pose in the operational space is used for the dynamic identification of parallel robots. A modified Finite Fourier series-based exciting trajectory for parallel robots is proposed as:

$$\chi_i(t) = \sum_{l=1}^{n} [\frac{sin(2\pi\omega_0 lt)}{2\pi\omega_0 l} s_i^l - \frac{cos(2\pi\omega_0 lt)}{2\pi\omega_0 l} c_i^l] + \chi_{0_i} \tag{37}$$

where $\chi_i(t)$ is the ith column of the pose trajectory. Therefore $2n+1$ parameters $\delta_i = [s_i^1, c_i^1, \cdots, s_i^n, c_i^n, \chi_{0_i}]^T$ can be estimated by solving a nonlinear optimization problem. The singularity check should be implemented for each sampled pose. According to previous research work [31,41], the maximum wrench rotation range inside the singularity-free domain of the 6-RSS parallel robot is (−0.9948 rad, 0.9948 rad). The inverse kinematic model is used to map the poses into the joint space and to check if the joint angles stay inside the singularity-free domain. To obtain the operational space states in the regression matrix H, the time derivatives of the Euler-angle should be converted to the angular velocity and acceleration, as shown in Equations (2) and (38).

$$\dot{\omega} = \begin{bmatrix} \ddot{\alpha} + \ddot{\gamma} s\beta + \dot{\beta}\dot{\gamma} c\beta \\ c\alpha(\ddot{\beta} - \dot{\alpha}\dot{\gamma} c\beta) + s\alpha(\dot{\beta}\dot{\gamma} s\beta - \dot{\beta}\dot{\alpha} - \ddot{\gamma} c\beta) \\ s\alpha(\ddot{\beta} - \dot{\alpha}\dot{\gamma} c\beta) + c\alpha(\dot{\beta}\dot{\alpha} + \ddot{\gamma} c\beta - \dot{\beta}\dot{\gamma} s\beta) \end{bmatrix}. \tag{38}$$

As shown in Equation (26), the observability index of H should be maximized to achieve a good identification result for given the exciting trajectories. The observability index O_{in} used in [31] is chosen as the criteria. Therefore the optimal exciting trajectory can be obtained by solving the following nonlinear optimization problem:

$$Maximize\ \ O_{in}(\delta) = \frac{\sqrt[\varsigma]{\sigma_1 \sigma_2 \cdots \sigma_\varsigma}}{\sqrt{m}} = \frac{\sqrt[\varsigma]{det(\sqrt{H^T H})}}{\sqrt{m}} \tag{39}$$

where the singular values of H are denoted by $\sigma_1 \geq \sigma_2 \geq \cdots \geq \sigma_\varsigma$, m is the number of sampled poses of the trajectory, ς is the number of dynamic parameters to be identified.

3.4. The Procedure of Identification

The whole procedure of the proposed closed-loop output-error identification method is given in Figure 4. Firstly, the dynamic model of the parallel robot is derived as Equations (17) and (25). Then the optimized exciting trajectory, $\chi_d(t)$, can be generated by using the method mentioned in previous subsection. By using $\chi(t)$ as reference input signal to the outer loop visual servoing control systems of both the real plant and the simulated model, the measured output pose $\chi_m(t)$ of the parallel robot by the optical CMM sensor is compared with that of the simulated model. The identification of the parameters $\boldsymbol{\Lambda}$ is carried out by solving the nonlinear optimization problem. Lastly, the identified model can be validated by feeding several testing trajectories to the systems.

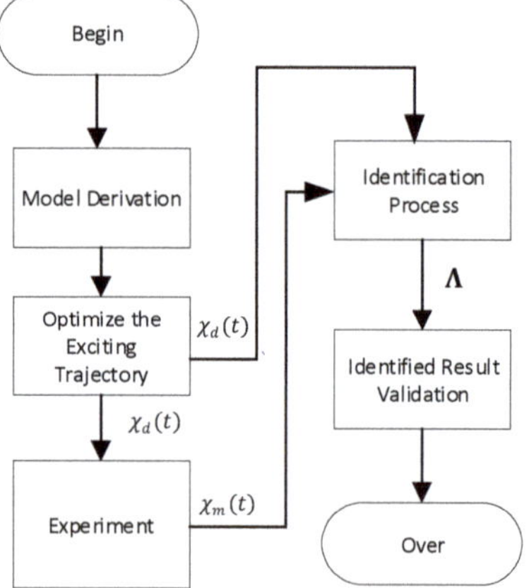

Figure 4. Sketch of the identification procedure.

4. Simulation and Experiment Results

In this section, the dynamic model is verified by the simulation using Matlab/SimMechanics. In addition, the closed-loop output-error identification is carried out on a 6-RSS parallel robot. An outer loop visual servoing controller is implemented on the real plant and the simulated model individually. The C-track 780 from Creaform Inc. is adopted to measure the pose of the end-effector of parallel robot.

4.1. Model Validation

The analytical dynamic model is rather complex and it is a non-trival task to ensure that the code of dynamic model is mistake free. A simulation verification method is used to verify the built mathematical dynamic model. A mechanical model of the 6-RSS parallel robot is built by using the Multibody SimMechanics Toolbox of Matlab/Simulink. The SimMechanics model is constructed by choosing the parts from SimMechanics library as shown in Figure 5. The coordinates frames and physical parameters of rigid body blocks can be specified in the setting option. The revolute and spherical joints are used to connect the rigid bodies. The body sensor part can provide the position and orientation of the coordinate frames. The entire model is directly actuated by the torque of the motors, and 3D animation shown in Figure 6 can be provided by SimMechanics. It should be noted that SimMechanics model can only simulate the forward dynamics and cannot be used for controller design. However, it is relatively easy and intuitive to build SimMechanics model with high fidelity [42]. To validate the mathematical dynamic model Equation (17), the explicit form of the forward dynamic model obtained by Equation (40) is employed to compare with the SimMechanics model.

$$\dot{v}_E = M(\chi_E)^{-1}(\tau_g - C(\chi_E, v_E)v_E - G(\chi_E) - \tau_f), \tag{40}$$

The mathematical dynamic model is built by using S-function of Simulink. The initial dynamic model parameters of both SimMechanics and mathematical models are derived from manufacturer specifications and are given in Table 1.

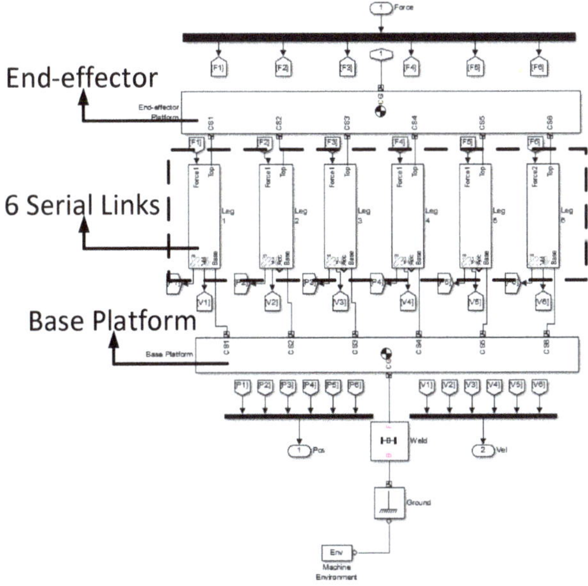

Figure 5. Mechanical model of 6-RSS parallel robot built by SimMechanics.

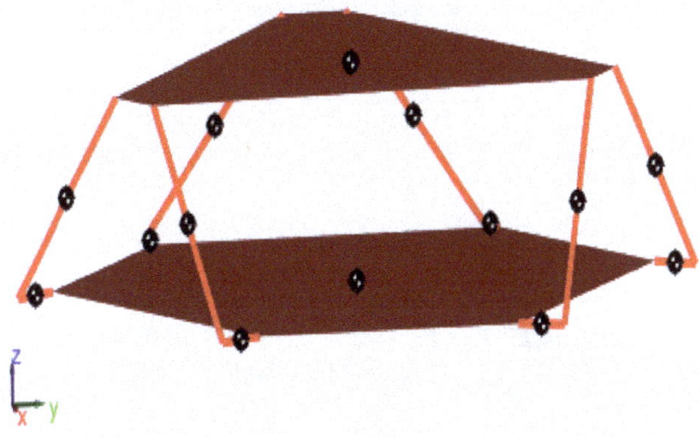

Figure 6. 3D animation of 6-RSS parallel robot. https://youtu.be/HXtCvgkn2jw.

Table 1. Initial dynamic model parameters of 6-RSS parallel robot.

Dynamic Model Parameters	Initial Value
m_p (kg)	24.0
I_{x_p} 10^{-2} (kg · m^2)	17.8
I_{y_p} 10^{-2} (kg · m^2)	17.8
I_{z_p} 10^{-2} (kg · m^2)	35.0
m_{wi} 10^{-2} (kg)	68.5
$I_{x_{w_i}}$ 10^{-5} (kg · m^2)	22.9
$I_{y_{w_i}}$ 10^{-5} (kg · m^2)	22.9
$I_{z_{w_i}}$ 10^{-5} (kg · m^2)	60.5
m_{li} (kg)	1.31
$I_{x_{l_i}}$ 10^{-5} (kg · m^2)	52.3
$I_{y_{l_i}}$ 10^{-5} (kg · m^2)	52.3
$I_{z_{l_i}}$ 10^{-4} (kg · m^2)	21.3

As shown in Figure 7, a simple PID controller is used to stabilize both mathematical and SimMechanic models with the same PID gains. The exciting trajectory $\chi_d(t)$ derived from Equation (36) is used as the reference input signal. The outputs of the SimMechanics and mathematical dynamic model (Equation (40)) are $\chi_{\_s}(t)$ and $\chi_{\_m}(t)$ respectively. The difference between $\chi_{\_s}(t)$ and $\chi_{\_m}(t)$ is shown in Figure 8. The maximum position and angle errors are around 1.25 mm and 3.25×10^{-3} rad, which occur at the beginning of the simulation, and are often caused by the kinematic error. The largest steady-state errors are about 0.1 mm in the position and 0.25×10^{-3} rad in the angle, which can prove the correctness of the mathematical model. The validated mathematical model can be used in the simulation part for the subsequent identification.

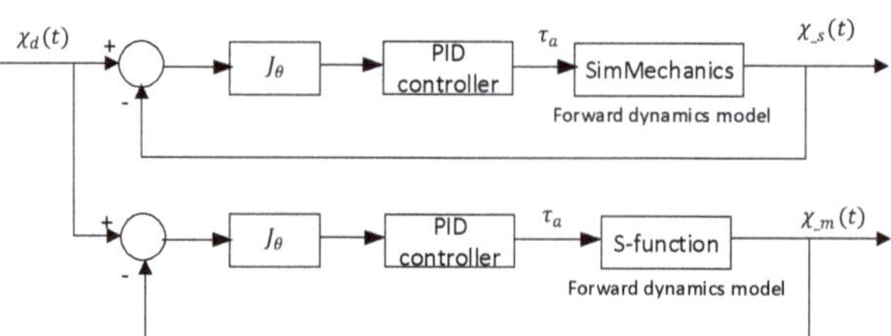

Figure 7. Dynamic model verification block diagram of 6-RSS parallel robot.

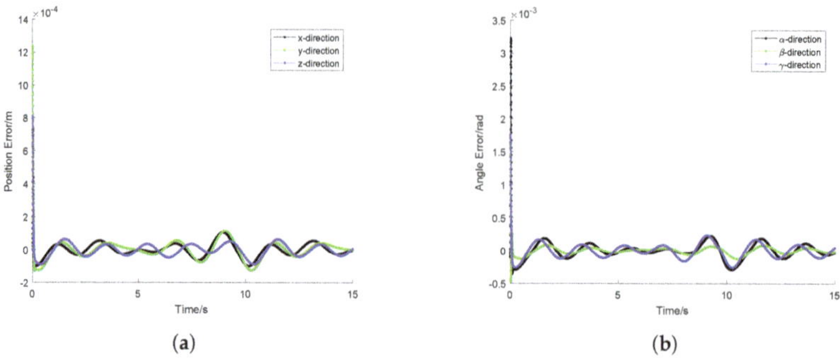

Figure 8. Simulation results of SimMechanics model. (**a**) Position error; (**b**) Angle error.

4.2. Identification Experiment

The experiment setup is shown in Figure 9. The 6-RSS parallel robot with 6 built-in joint controllers is provided by Servo & Simulation Inc. (Sanford, FL, USA). The built-in controllers communicate with the robot control computer through two Quanser MultiQ-PCI (Sensoray Model 626) data acquisition cards provided by Quanser Inc. (Markham, ON, Canada). Quanser's QUARC^{TM} software is running on the robot control computer with Windows 7.0 32-bit operating system and Intel Core Processor i7-3770 3.4 GHz. QUARC^{TM} software is capable of generating real-time application though Simulink based controllers and implementing the application in real time on the Windows target. The C-track 780 provided by Creaform Inc. (Levis, QC, Canada) is used to obtain the image data of the reflectors attached on the robot. The reflectors provided by Creaform Inc. are magnetic stickers which are easily fixed on the robots and are used as the feature points. In another Windows 7.0 64-bit computer with Intel Xeon Processor E5-1650 v3 3.5 GHz and NVIDIA Quadro K2200 (Santa Clara, CA, USA) professional graphics board, Vxelements software is used to process the image data and transmit the pose of the end-effector to the robot control computer.

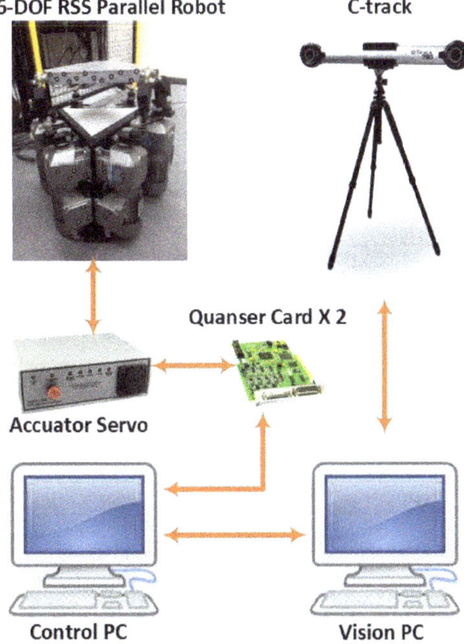

Figure 9. Architecture of the Experiment System.

As shown in Figure 10, the reflectors are stuck on the surface of the moving platform. At least three non-collinear points on each plane of Plane A, Plane B and Plane C are employed to build up the equations of planes based on Cramer's rule. Then the intersection lines and points of three planes can be used to define the x direction of ΣE, and z direction is aligned with the norm of Plane A. The origin point of ΣE is derived from the intersection point of l_1 and l_2. Then, the obtained ΣE in the optical CMM sensor frame is directly used as the target frame. The base frame of the parallel robot is defined by following the similar procedure.

To eliminate the high frequency noise of the pose measurement from the optical CMM, measurement data is filtered by the zero-phase forward and reverse 8th order Butterworth filter with the cut-off frequency 60 Hz. The filtering process is carried out by the Zero-phase digital filtering function of Maltab, $filtfilt$.

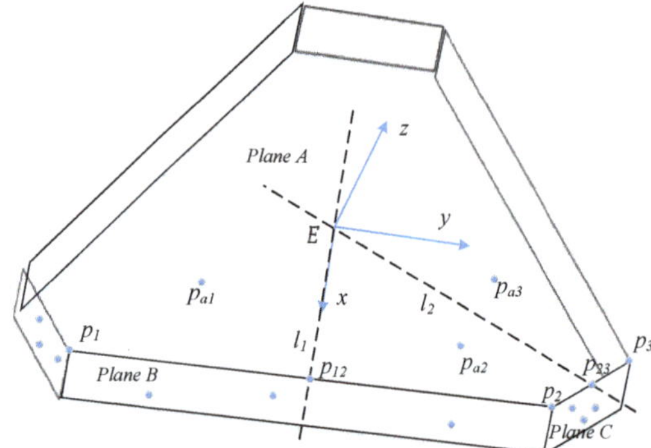

Figure 10. Measurement of ΣE.

The optimization of the exciting trajectory is carried out by using the GA function of the Optimization Toolbox of Matlab. The GA algorithm uses binary code to represent the harmonic parameters. The fundamental frequency f_0 is selected as 0.1 Hz and the harmonics number n is chosen as 5. By taking the observation index as the fitness function, the binary code is updated to maximize the fitness value in each step. The starting value of the observation index O_{in} is 0.38 and the stop criteria is set as 10^{-10}. The algorithm stops after 324 iterations with the maximum O_{in} 1.345. The derived optimal exciting trajectory is given in Figure 11.

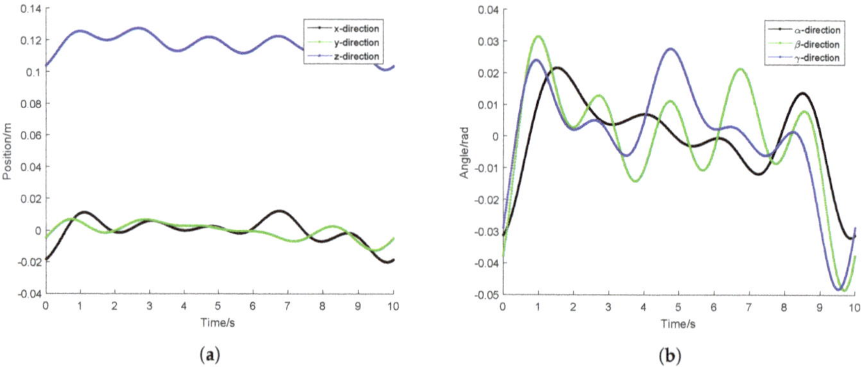

Figure 11. Optimal exciting trajectory. (**a**) Positional trajectory; (**b**) Angular trajectory.

The identification procedure is implemented off-line through simulation illustrated in Section 3.2. The optimization procedure is carried out by using optimization functions in Matlab R2016a. The minimal performance requirements of the computation platform are given as: 4 cores Inter or AMD Processor; 6 GB disk space; 4 GB RAM. The same outer loop visual servoing controller, Equation (30), is employed in the simulation. The gains of the visual servoing controller are obtained through trial and error in the extensive experimental tests. The well-tuned gains are given as $k_p = 0.3, k_d = 0.001, k_i = 2.4$. and the built-in joint PID controllers, Equation (31), are also implemented in the simulation. The dynamic parameters derived from manufacturer specifications are used as initial values, and the initial values of the PID gains in Equation (31) are obtained through trial and error based on the simulation model. During the tuning, in the simulation system, the inertial and friction parameters are set as the values based on the manufacturer specifications, and the visual servoing controller gains are set as the same values of the controller of the real system. The identified parameters, given in Table 2, are derived by using $lsqnonlin$ function of the Optimization Toolbox of Matlab after 8 iterations. A total of 50 out of 67 parameters are identified, and are used in the simulation model, Equation (40), to capture the dynamic characteristic of the parallel robot. The other parameters either do not contribute to or have slight impact on the dynamics of the parallel robot. Those parameters can be eliminated by using QR decomposition on the regression matrix H [23,43]. If any diagonal elements of $\underline{R}$ are smaller than the pre-defined small number, i.e., $R_{ii} < \varepsilon$, where ε is chosen as 10^{-3} in the paper, the corresponding columns of the regression matrix H are deleted. By doing so, the matrix H is better conditioned and the identification procedure is sped up.

Then, the pose trajectories are generated by using the identified parameters in the simulation, and are compared with the pose measurement, as shown in Figure 12. Table 3 shows the root-mean-square (RMS) levels of the pose trajectory errors.

Table 2. Identified parameters of 6-RSS parallel robot.

Parameters	Initial Value	Identified Value	Parameters	Initial Value	Identified Value
m_p (kg)	24.0	23.5	m_{l4} (kg)	1.31	1.24
I_{x_p} 10^{-2} (kg · m^2)	17.8	16.1	$I_{x_{l_4}}$ 10^{-5} (kg · m^2)	52.3	12.9
I_{y_p} 10^{-2} (kg · m^2)	17.8	14.4	$I_{y_{l_4}}$ 10^{-5} (kg · m^2)	52.3	39.2
I_{z_p} 10^{-2} (kg · m^2)	35.0	33.7	m_{l5} (kg)	1.31	1.33
m_{w1} 10^{-2} (kg)	68.5	67.1	$I_{x_{l_5}}$ 10^{-5} (kg · m^2)	52.3	30.7
$I_{z_{w_1}}$ 10^{-5} (kg · m^2)	60.5	69.4	$I_{y_{l_5}}$ 10^{-5} (kg · m^2)	52.3	49.7
m_{w2} 10^{-2} (kg)	68.5	68.7	m_{l6} (kg)	1.31	1.34
$I_{z_{w_2}}$ 10^{-5} (kg · m^2)	60.5	10.0	$I_{x_{l_6}}$ 10^{-5} (kg · m^2)	52.3	47.0
m_{w3} (kg) 10^{-2} (kg)	68.5	68.0	$I_{y_{l_6}}$ 10^{-5} (kg · m^2)	52.3	50.9
$I_{z_{w_3}}$ 10^{-5} (kg · m^2)	60.5	22.6	f_{c_1}	0	0.104
m_{w4} 10^{-2} (kg)	68.5	67.7	f_{v_1}	0	0.148
$I_{z_{w_4}}$ 10^{-5} (kg · m^2)	60.5	76.0	f_{c_2}	0	0.111
m_{w5} (kg) 10^{-2} (kg)	68.5	68.3	f_{v_2}	0	0.187
$I_{z_{w_5}}$ 10^{-5} (kg · m^2)	60.5	44.1	f_{c_3}	0	0.0336
m_{w6} (kg) 10^{-2} (kg)	68.5	68.4	f_{v_3}	0	0.0993
$I_{z_{w_6}}$ 10^{-5} (kg · m^2)	60.5	69.6	f_{c_4}	0	0.147
m_{l1} (kg)	1.31	1.33	f_{v_4}	0	0.854
$I_{x_{l_1}}$ 10^{-5} (kg · m^2)	52.3	63.0	f_{c_5}	0	0.104
$I_{y_{l_1}}$ 10^{-5} (kg · m^2)	52.3	68.4	f_{v_5}	0	0.0803
m_{l2} (kg)	1.31	1.32	f_{c_6}	0	0.0828
$I_{x_{l_2}}$ 10^{-5} (kg · m^2)	52.3	60.7	f_{v_6}	0	0.0349
$I_{y_{l_2}}$ 10^{-5} (kg · m^2)	52.3	71.8	l_p	10	10.5
m_{l3} (kg)	1.31	1.25	l_i	12	11.4
$I_{x_{l_3}}$ 10^{-5} (kg · m^2)	52.3	22.4	l_d	0.1	0.164
$I_{y_{l_3}}$ 10^{-5} (kg · m^2)	52.3	35.5			

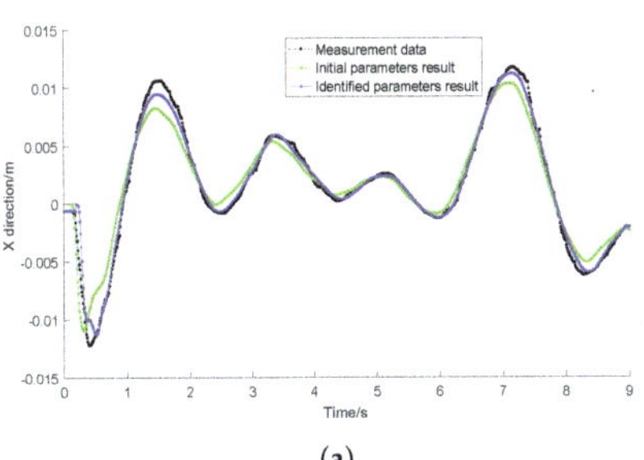

(a)

(b)

Figure 12. *Cont.*

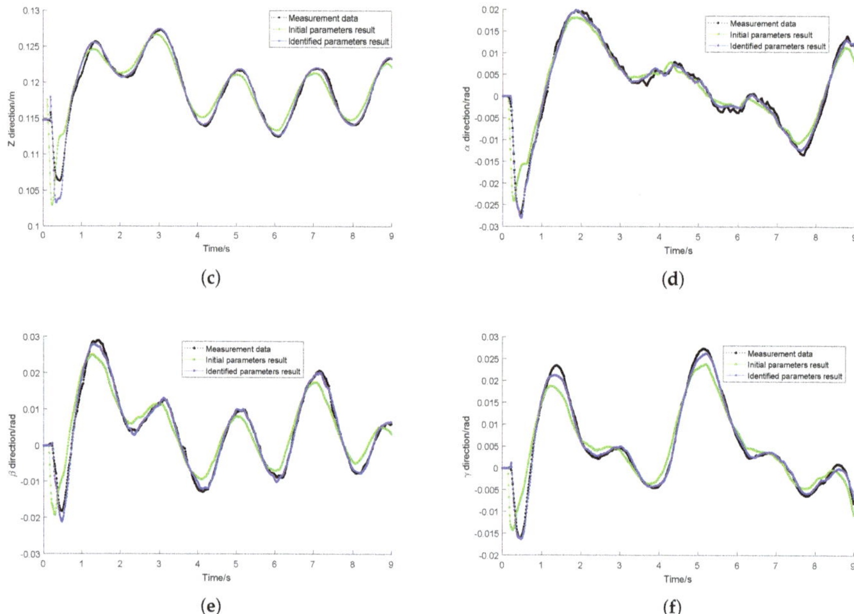

Figure 12. The pose trajectories of the parallel robot: the measurement of the real plant (**black dot**), the output of the simulation with initial parameters (**green line**), the output of the simulation with identified parameters (**blue line**), (**a**) Along X Direction; (**b**) Along Y Direction; (**c**) Along Z Direction; (**d**) Around α Axis; (**e**) Around β Axis; (**f**) Around γ Axis.

Table 3. The RMS levels of the pose trajectory errors.

	Before Identification	After Identification
x direction (mm)	1.26	0.408
y direction (mm)	1.16	0.235
z direction (mm)	1.55	0.494
α direction 10^{-3} (rad)	2.52	0.956
β direction 10^{-3} (rad)	3.58	0.797
γ direction 10^{-3} (rad)	2.80	0.725

4.3. *Identified Results Validation*

To verify the identified parameters, ten more trajectories are generated according to Equation (36) with random harmonic parameters under the singularity constraint. The generated trajectories are used as desired trajectories, and are fed to the parallel robot and the identified model in the simulation respectively. The RMS levels of the pose trajectory errors are given in Table 4, and the measurement and the simulated pose trajectories are given in Figure 13 according to the 1st desired trajectory. The RMS of the position and orientation errors for all ten trajectories are below 0.8 mm and 1.4×10^{-3} rad respectively, which validate the identified results of previous subsection. In addition, the proposed identification procedure is implemented based on the ten trajectories to analyze the statistic property of the identification results. After deriving ten more groups of identified parameters, the variation measure of the identification results are given in Table 5. The highest relative variation of the parameter is below 25%, which is acceptable. It has been stated that less than 30 percent in the variation measure of the parameters gives a good match to the real system [44].

Table 4. The RMS levels of the ten validation trajectory errors.

	x (mm)	y (mm)	z (mm)	α 10^{-3} (rad)	β 10^{-3} (rad)	γ 10^{-3} (rad)
1st	0.517	0.384	0.709	1.091	0.963	0.966
2nd	0.273	0.410	0.522	1.157	1.071	0.932
3rd	0.381	0.403	0.600	1.242	0.830	1.143
4th	0.341	0.511	0.617	1.136	0.814	1.161
5th	0.322	0.394	0.464	1.219	0.877	0.835
6th	0.310	0.301	0.441	1.195	1.111	1.040
7th	0.360	0.402	0.455	1.263	1.176	1.024
8th	0.418	0.483	0.460	1.251	1.211	1.015
9th	0.342	0.510	0.557	1.411	1.156	0.711
10th	0.318	0.379	0.473	1.219	1.023	0.905

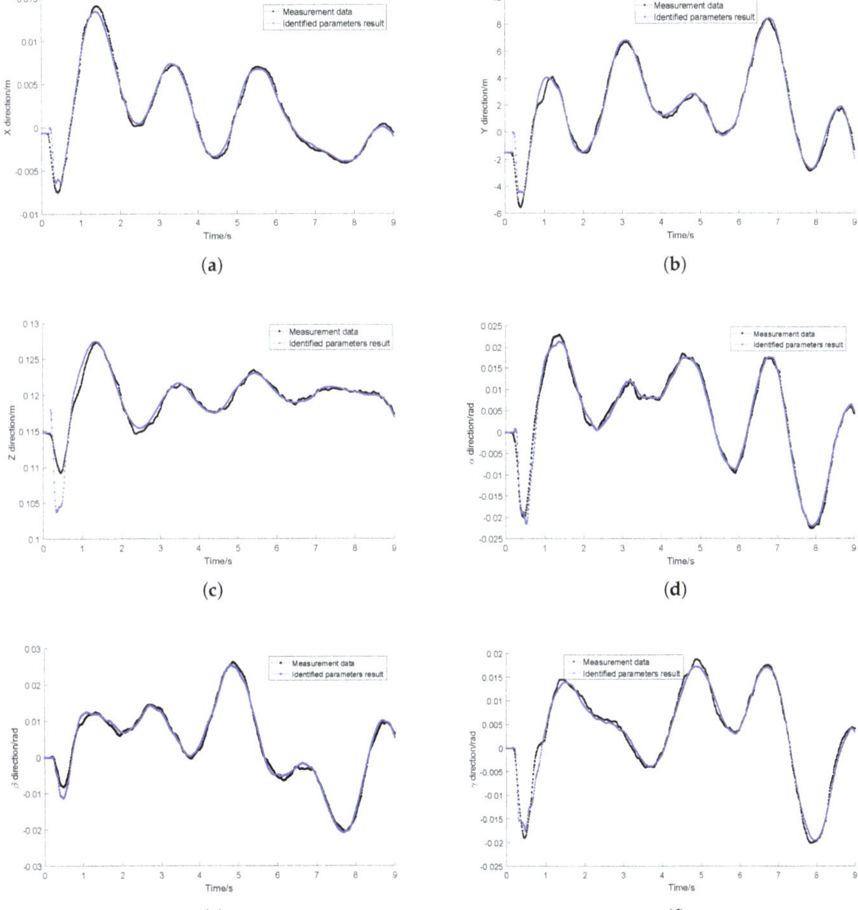

Figure 13. The pose trajectories of the parallel robot: the measurement of the real plant (**black dot**), the output of the simulation with identified parameters (**blue line**), (**a**) Along X Direction; (**b**) Along Y Direction; (**c**) Along Z Direction;(**d**) Around α Axis; (**e**) Around β Axis; (**f**) Around γ Axis.

Table 5. Variation measure of the identification result.

Parameters	Variation Measure	Parameters	Variation Measure
m_p (kg)	0.48%	m_{l4} (kg)	19.1%
I_{x_p} (kg · m^2)	5.0%	$I_{x_{l_4}}$ (kg · m^2)	0.16%
I_{y_p} (kg · m^2)	24.9%	$I_{y_{l_4}}$ (kg · m^2)	0.06%
I_{z_p} (kg · m^2)	8.6%	m_{l5} (kg)	10.1%
m_{w1} (kg)	0.56%	$I_{x_{l_5}}$ (kg · m^2)	0.15%
$I_{z_{w_1}}$ (kg · m^2)	6.6%	$I_{y_{l_5}}$ (kg · m^2)	0.01%
m_{w2} (kg)	1.5%	m_{l6} (kg)	12.9%
$I_{z_{w_2}}$ (kg · m^2)	1.4%	$I_{x_{l_6}}$ (kg · m^2)	0.17%
m_{w3} (kg)	3.4%	$I_{y_{l_6}}$ (kg · m^2)	0.02%
$I_{z_{w_3}}$ (kg · m^2)	5.0%	f_{c_1}	15.9%
m_{w4} (kg)	3.7%	f_{v_1}	11.7%
$I_{z_{w_4}}$ (kg · m^2)	3.8%	f_{c_2}	21.4%
m_{w5} (kg)	6.6%	f_{v_2}	13.9%
$I_{z_{w_5}}$ (kg · m^2)	6.6%	f_{c_3}	9.0%
m_{w6} (kg)	0.79%	f_{v_3}	16.4%
$I_{z_{w_6}}$ (kg · m^2)	1.7%	f_{c_4}	17.8%
m_{l1} (kg)	4.1%	f_{v_4}	15.2%
$I_{x_{l_1}}$ (kg · m^2)	0.04%	f_{c_5}	15.1%
$I_{y_{l_1}}$ (kg · m^2)	0.11%	f_{v_5}	21.8%
m_{l2} (kg)	5.2%	f_{c_6}	23.4%
$I_{x_{l_2}}$ (kg · m^2)	0.05%	f_{v_6}	13.0%
$I_{y_{l_2}}$ (kg · m^2)	0.12%	l_p	2.82%
m_{l3} (kg)	20.9%	l_i	0.44%
$I_{x_{l_3}}$ (kg · m^2)	0.09%	l_d	3.7%
$I_{y_{l_3}}$ (kg · m^2)	0.17%		

Therefore by using the proposed visual closed-loop output-error identification method, the identified dynamic model can approximate the real plant with acceptable accuracy.

5. Conclusions and Further Works

In this paper, a visual closed-loop output-error identification method based on an optical CMM sensor for parallel robots is proposed. An outer loop visual servoing controller is employed in both the real plant and the simulation model to stabilize the two systems. The benefits of the proposed method are summarized as follows: elimination of the need for the joint and torque measurements, the exact knowledge of the built-in joint controller of the industrial robots, and the time-consuming forward kinematics calculation. The correctness and accuracy of the built dynamic model are validated by the Matlab/SimMechanics simulation. The experimental test results show that the identified dynamic model can capture the dynamics of the real parallel robot with satisfactory accuracy. The proposed method can be easily applied to other types of industrial parallel robots with unknown PID built-in controller or its variant, such as 6 DOF Stewart platforms, 6 UPS and 6 RUS parallel robots etc. The complexity of those dynamic models is similar to that of the 6-RSS parallel robot. Since the analytical solution of the forward kinematics of those 6 DOF parallel robots does not exist, the proposed visual identification method does not need the forward kinematic model and hence has a lot of advantages. The proposed identification method can also be applied to parallel robots with less DOF than 6 DOF. Taking the advantages of the visual sensor, the dynamic model can be identified for the visual servoing purpose. In the future, the advanced model-based visual servoing control method will be further studied to improve the tracking performance of parallel robots based on the identification results.

Author Contributions: The authors' individual contributions are provided as follow: conceptualization, P.L., A.G., W.X. and W.T.; methodology, P.L., A.G., W.X. and W.T.; software, P.L.; validation, P.L., A.G., and W.X.; resources, W.X.; writing–original draft preparation, P.L.; writing–review and editing, P.L., A.G., W.X. and W.T.; supervision, W.X. and W.T.; project administration, W.X.; funding acquisition, W.X.

Funding: This research was funded by Natural Sciences and Engineering Research Council of Canada grant number N00892.

Conflicts of Interest: The authors declare no conflict of interest. The funder had no role in the design of the study; in the collection, analyses, or interpretation of data; in the writing of the manuscript, or in the decision to publish the results.

Appendix A

The derivative of the wrench Jacobian, $\dot{J}_{a_i}$, is given as follows:

$$\dot{J}_{a_i} = \begin{bmatrix} \dot{J}_{au_i} \\ \hat{\mathbf{s}}\dot{J}_{ad_i} \end{bmatrix}, \tag{A1}$$

where

$$\dot{J}_{ad_i} = -\frac{\dot{m}}{m^2}\begin{bmatrix} \boldsymbol{l}_i^T & (\boldsymbol{a}_i \times \boldsymbol{l}_i)^T \end{bmatrix} + \frac{1}{m}\begin{bmatrix} (\boldsymbol{\omega}_{2i} \times \boldsymbol{l}_i)^T & ([\boldsymbol{a}_i]_X[\boldsymbol{\omega}_2]_X\boldsymbol{l}_i)^T - ([\boldsymbol{l}_i]_X[\boldsymbol{\omega}]_X\boldsymbol{a}_i)^T \end{bmatrix}, \tag{A2}$$

and

$$\begin{aligned} m &= (\boldsymbol{w}_i \times \boldsymbol{l}_i) \cdot \hat{\mathbf{s}} \\ \dot{m} &= ([\boldsymbol{w}_i]_X[\boldsymbol{\omega}_2]_X\boldsymbol{l}_i \quad [\boldsymbol{l}_i]_X[\boldsymbol{\omega}_1]_X\boldsymbol{w}_i) \; \hat{\mathbf{s}} \\ \dot{J}_{au_i} &= ([\hat{\mathbf{s}}]_X[\boldsymbol{\omega}_1]_X\boldsymbol{c}_{w_i})J_{ad_i} + ([\hat{\mathbf{s}}]_X\boldsymbol{c}_{w_i})\dot{J}_{ad_i} \end{aligned} \tag{A3}$$

In addition, the link Jacobian $\dot{J}_{b_i}$ is obtained by:

$$\dot{J}_{b_i} = \begin{bmatrix} \dot{J}_{bu_i} \\ \dot{J}_{bd_i} \end{bmatrix} \tag{A4}$$

in which

$$\begin{aligned} \dot{J}_{bd_i} = &\frac{1}{\|\boldsymbol{l}_i\|^2}\{[\boldsymbol{\omega}_2 \times \boldsymbol{l}_i]_X[\boldsymbol{w}_i]_X\hat{\mathbf{s}}J_{ad_i} + \\ &\begin{bmatrix} [\boldsymbol{\omega}_2 \times \boldsymbol{l}_i]_X & [\boldsymbol{l}_i]_X[\boldsymbol{\omega} \times \boldsymbol{a}_i]_X - [\boldsymbol{\omega}_2 \times \boldsymbol{l}_i]_X[\boldsymbol{a}_i]_X \end{bmatrix} \\ &+ [\boldsymbol{l}_i]_X[\boldsymbol{\omega}_1 \times \boldsymbol{w}_i]_X\hat{\mathbf{s}}J_{ad_i} + [\boldsymbol{l}_i]_X[\boldsymbol{w}_i]_X\hat{\mathbf{s}}\dot{J}_{ad_i}\}, \end{aligned} \tag{A5}$$

and

$$\dot{J}_{bu_i} = -[\boldsymbol{\omega}_1 \times \boldsymbol{w}_i]_X\hat{\mathbf{s}}J_{ad_i} - [\boldsymbol{w}_i]_X\hat{\mathbf{s}}\dot{J}_{ad_i} - [\boldsymbol{\omega}_2 \times \boldsymbol{l}_i]_X J_{bd_i} - [\boldsymbol{c}_{l_i}]_X\dot{J}_{bd_i} \tag{A6}$$

References

1. Miletović, I.; Pool, D.; Stroosma, O.; van Paassen, M.; Chu, Q. Improved Stewart platform state estimation using inertial and actuator position measurements. *Control Eng. Pract.* **2017**, *62*, 102–115. [CrossRef]
2. Abdellatif, H.; Heimann, B. Model-Based Control for Industrial Robots: Uniform Approaches for Serial and Parallel Structures. In *Industrial Robotics*; Cubero, S., Ed.; IntechOpen: Rijeka, Croatia, 2006; Chapter 19; pp. 523–556, doi:10.5772/5037.
3. Bai, S.; Teo, M.Y. Kinematic calibration and pose measurement of a medical parallel manipulator by optical position sensors. *J. Robot. Syst.* **2003**, *20*, 201–209. [CrossRef]
4. Absolute Accuracy Industrial Robot Option. 2011. Available online: https://www.google.com.hk/url?sa=t&rct=j&q=&esrc=s&source=web&cd=2&ved=2ahUKEwiK6L_fndLjAhUQTY8KHYTaBjcQFjABegQIChAE&url=https%3A%2F%2Flibrary.e.abb.com%2Fpublic%2F0f879113235a0e1dc1257b130056d133%2FAbsolute%2520Accuracy%2520EN_R4%2520US%252002_05.pdf&usg=AOvVaw2YHKD4M3X8PjRHuBDV8y_F (accessed on 7 November 2018).

5. Taghirad, H.D. *Parallel Robots: Mechanics and Control*; CRC Press: Boca Raton, FL, USA, 2013.
6. Merlet, J.P. *Parallel Robots*; Springer Science & Business Media: Berlin, Germany, 2012.
7. Dasgupta, B.; Mruthyunjaya, T. A Newton-Euler formulation for the inverse dynamics of the Stewart platform manipulator. *Mech. Mach. Theory* **1998**, *33*, 1135–1152. [CrossRef]
8. Nguyen, C.C.; Pooran, F.J. Dynamic analysis of a 6 DOF CKCM robot end-effector for dual-arm telerobot systems. *Robot. Auton. Syst.* **1989**, *5*, 377–394. [CrossRef]
9. Wang, J.; Gosselin, C.M. A new approach for the dynamic analysis of parallel manipulators. *Multibody Syst. Dyn.* **1998**, *2*, 317–334. [CrossRef]
10. Zanganeh, K.E.; Sinatra, R.; Angeles, J. Kinematics and dynamics of a six-degree-of-freedom parallel manipulator with revolute legs. *Robotica* **1997**, *15*, 385–394. [CrossRef]
11. Dumlu, A.; Erenturk, K. Modeling and trajectory tracking control of 6-DOF RSS type parallel manipulator. *Robotica* **2014**, *32*, 643–657. [CrossRef]
12. Ljung, L. (Ed.) *System Identification (2nd ed.): Theory for the User*; Prentice Hall PTR: Upper Saddle River, NJ, USA, 1999.
13. Hollerbach, J.; Khalil, W.; Gautier, M. Model identification. In *Springer Handbook of Robotics*; Springer: Berlin/Heidelberg, Germany, 2016; pp. 113–138.
14. Swevers, J.; Verdonck, W.; De Schutter, J. Dynamic model identification for industrial robots. *IEEE Control Syst. Mag.* **2007**, *27*, 58–71.
15. Wu, J.; Wang, J.; You, Z. An overview of dynamic parameter identification of robots. *Robot. Comput.-Integr. Manuf.* **2010**, *26*, 414–419. [CrossRef]
16. Calanca, A.; Capisani, L.M.; Ferrara, A.; Magnani, L. MIMO closed loop identification of an industrial robot. *IEEE Trans. Control Syst. Technol.* **2010**, *19*, 1214–1224. [CrossRef]
17. Olsen, M.M.; Swevers, J.; Verdonck, W. Maximum likelihood identification of a dynamic robot model: Implementation issues. *Int. J. Robot. Res.* **2002**, *21*, 89–96. [CrossRef]
18. Janot, A.; Vandanjon, P.O.; Gautier, M. An instrumental variable approach for rigid industrial robots identification. *Control Eng. Pract.* **2014**, *25*, 85–101. [CrossRef]
19. Gautier, M.; Janot, A.; Vandanjon, P.O. A new closed-loop output error method for parameter identification of robot dynamics. *IEEE Trans. Control Syst. Technol.* **2013**, *21*, 428–444. [CrossRef]
20. Walter, E.; Pronzato, L. *Identification of Parametric Models from Experimental Data*; Springer: Berlin/Heidelberg, Germany, 1997.
21. Del Prete, A.; Mansard, N.; Ramos, O.E.; Stasse, O.; Nori, F. Implementing torque control with high-ratio gear boxes and without joint-torque sensors. *Int. J. Humanoid Robot.* **2016**, *13*, 1550044. [CrossRef]
22. Brunot, M.; Janot, A.; Young, P.C.; Carrillo, F. An improved instrumental variable method for industrial robot model identification. *Control Eng. Pract.* **2018**, *74*, 107–117. [CrossRef]
23. Briot, S.; Gautier, M. Global identification of joint drive gains and dynamic parameters of parallel robots. *Multibody Syst. Dyn.* **2015**, *33*, 3–26. [CrossRef]
24. Grotjahn, M.; Heimann, B.; Abdellatif, H. Identification of friction and rigid-body dynamics of parallel kinematic structures for model-based control. *Multibody Syst. Dyn.* **2004**, *11*, 273–294. [CrossRef]
25. Abdellatif, H.; Heimann, B. Advanced model-based control of a 6-DOF hexapod robot: A case study. *IEEE/ASME Trans. Mechatron.* **2010**, *15*, 269–279. [CrossRef]
26. Vivas, A.; Poignet, P.; Marquet, F.; Pierrot, F.; Gautier, M. Experimental dynamic identification of a fully parallel robot. In Proceedings of the 2003 IEEE International Conference on Robotics and Automation (Cat. No. 03CH37422), Taipei, Taiwan, 14–19 September 2003; IEEE: Piscataway, NJ, USA, 2003; Volume 3, pp. 3278–3283.
27. Guegan, S.; Khalil, W.; Lemoine, P. Identification of the dynamic parameters of the orthoglide. In Proceedings of the 2003 IEEE International Conference on Robotics and Automation (Cat. No. 03CH37422), Taipei, Taiwan, 14–19 September 2003; IEEE: Piscataway, NJ, USA, 2003; Volume 3, pp. 3272–3277.
28. Renaud, P.; Vivas, A.; Andreff, N.; Poignet, P.; Martinet, P.; Pierrot, F.; Company, O. Kinematic and dynamic identification of parallel mechanisms. *Control Eng. Pract.* **2006**, *14*, 1099–1109. [CrossRef]
29. Lightcap, C.; Banks, S. Dynamic identification of a mitsubishi pa10-6ce robot using motion capture. In Proceedings of the 2007 IEEE/RSJ International Conference on Intelligent Robots and Systems, San Diego, CA, USA, 29 October–2 November 2007; pp. 3860–3865.

30. Chellal, R.; Cuvillon, L.; Laroche, E. Model identification and vision-based H∞ position control of 6-DoF cable-driven parallel robots. *Int. J. Control* **2017**, *90*, 684–701. [CrossRef]
31. Li, P.; Zeng, R.; Xie, W.; Zhang, X. Relative posture-based kinematic calibration of a 6-RSS parallel robot by optical coordinate measurement machine. *Int. J. Adv. Robot. Syst.* **2018**, *15*, 1729881418765861. [CrossRef]
32. Shu, T.; Gharaaty, S.; Xie, W.; Joubair, A.; Bonev, I.A. Dynamic Path Tracking of Industrial Robots With High Accuracy Using Photogrammetry Sensor. *IEEE/ASME Trans. Mechatron.* **2018**, *23*, 1159–1170. [CrossRef]
33. Xie, W.; Krzeminski, M.; El-Tahan, H.; El-Tahan, M. Intelligent friction compensation (IFC) in a harmonic drive. In Proceedings of the IEEE NECEC, St. John's, NL, Canada, 13 November 2002.
34. Hartley, R.I.; Sturm, P. Triangulation. *Comput. Vis. Image Underst.* **1997**, *68*, 146–157. [CrossRef]
35. Wilson, W.J.; Hulls, C.C.W.; Bell, G.S. Relative end-effector control using cartesian position based visual servoing. *IEEE Trans. Robot. Autom.* **1996**, *12*, 684–696. [CrossRef]
36. Yuan, J.S. A general photogrammetric method for determining object position and orientation. *IEEE Trans. Robot. Autom.* **1989**, *5*, 129–142. [CrossRef]
37. Brosilow, C.; Joseph, B. *Techniques of Model-Based Control*; Prentice Hall Professional: Upper Saddle River, NJ, USA, 2002.
38. Marlin, T.E. Process Control, Designing Processes and Control Systems for Dynamic Performance. Available online: https://pdfs.semanticscholar.org/c068/dedc943743ce3ee416ba9202a7172b715e3a.pdf (accessed on 27 June 2019).
39. Park, K.J. Fourier-based optimal excitation trajectories for the dynamic identification of robots. *Robotica* **2006**, *24*, 625–633. [CrossRef]
40. Bona, B.; Curatella, A. Identification of industrial robot parameters for advanced model-based controllers design. In Proceedings of the 2005 IEEE International Conference on Robotics and Automation, Barcelona, Spain, 18–22 April 2005; pp. 1681–1686.
41. Zeng, R.; Dai, S.; Xie, W.; Zhang, X. Determination of the proper motion range of the rotary actuators of 6-RSS parallel robot. In Proceedings of the 2015 CCToMM Symposium on Mechanisms, Machines, and Mechatronics, Carleton University, Ottawa, ON, Canada, 28–29 May 2015; pp. 28–29.
42. Van Boekel, J. SimMechanics, MapleSim and Dymola: A First Look on Three Multibody Packages. 2009. Available online: https://www.semanticscholar.org/paper/SimMechanics-%2C-MapleSim-and-Dymola-%3A-a-first-look-Boekel/c51909107f1ddb67faf8bda1b387b0659ec04a7f (accessed on 7 November 2018).
43. Gautier, M. Numerical calculation of the base inertial parameters of robots. *J. Robot. Syst.* **1991**, *8*, 485–506. [CrossRef]
44. Taghirad, H.; Belanger, P. Modeling and parameter identification of harmonic drive systems. *J. Dyn. Syst. Meas. Control* **1998**, *120*, 439–444. [CrossRef]

Article

Optimal Image-Based Guidance of Mobile Manipulators Using Direct Visual Servoing

Álvaro Belmonte, José L. Ramón, Jorge Pomares *, Gabriel J. Garcia and Carlos A. Jara

Department of Physics, Systems Engineering and Signal Theory, University of Alicante, Alicante 03690, Spain; bb35@alu.ua.es (Á.B.); jl.ramon@ua.es (J.L.R.); gjgg@ua.es (G.J.G.); carlos.jara@ua.es (C.A.J.)

* Correspondence: jpomares@ua.es; Tel.: +34-965903400

Received: 20 February 2019; Accepted: 23 March 2019; Published: 27 March 2019

Abstract: This paper presents a direct image-based controller to perform the guidance of a mobile manipulator using image-based control. An eye-in-hand camera is employed to perform the guidance of a mobile differential platform with a seven degrees-of-freedom robot arm. The presented approach is based on an optimal control framework and it is employed to control mobile manipulators during the tracking of image trajectories taking into account robot dynamics. The direct approach allows us to take both the manipulator and base dynamics into account. The proposed image-based controllers consider the optimization of the motor signals sent to the mobile manipulator during the tracking of image trajectories by minimizing the control force and torque. As the results show, the proposed direct visual servoing system uses the eye-in-hand camera images for concurrently controlling both the base platform and robot arm. The use of the optimal framework allows us to derive different visual controllers with different dynamical behaviors during the tracking of image trajectories.

Keywords: visual servoing; optimal control; mobile manipulator; dynamic control

1. Introduction

This paper proposes a direct image-based visual servoing system for the guidance of a mobile manipulator. As described throughout the paper, the visual information is employed to guide both the robot manipulator and the base platform for the tracking of image trajectories. A common classification for visual servoing systems based on position and image is established [1]. Within the field of position-based visual servoing systems, it is worth mentioning some classic works [2] as well as some more recent ones such as the works of Huang et al. [3]. Besides these two works, it is worth mentioning other remarkable ones such as the works of Park et al. [4], which describes the use of vision for a robot's positioning with regard to an object whose position was previously determined. Position-based systems are those where the reference of the control system is in a desired tridimensional position and orientation. In contrast to position-based approaches, image-based visual servoing systems determine the reference in the image space. This strategy is the one used in this paper. Although the literature regarding image-based visual servoing is quite extensive, works, such as the presented in [1,5], perform a review of the classical approaches and problems related to indirect image-based controllers. One of the elements which most affects the behavior of image-based visual servoing controllers is the interaction matrix, which is employed by the controller and depends on several parameters. The interaction matrix J_s is employed by the image-based visual servoing controllers for relating velocities in a tridimensional point with velocities of a corresponding point in the image space. Deep learning approaches are currently in development for image-based control of robot manipulators (see e.g. [6,7]). When this kind of approaches has to be applied to a mobile manipulator, it is necessary to consider not only the manipulator but also the mobile platform, having a redundant system. When dealing with stability problems, it is worth mentioning those that stem from dealing with a redundant robot [8]. In this case, it is not only necessary to correctly perform the end effector guidance, but also to solve the redundancy by establishing the most appropriate joint configuration for each situation.

Another useful classification depending on the control strategy of visual servoing systems divides them in direct and indirect visual servoing. In indirect visual servoing, the control action is specified as velocities to be applied at the robot's end effector and does not take robot dynamics into account. In this case, the internal robot controller is used and translates these velocities into torques and forces for the joints. When applying an indirect controller to a task only kinematic aspects of the robot are taken into account. However, in direct controllers, the control action is usually given as forces and torques applied directly to the manipulator joints. By using this approach, the internal feedback control loop for the servomotors is removed, and the visual information is used to directly generate the currents or torques to be applied to the motors of the robot's every joint. Some direct visual control systems for redundant robots with chaos compensation have been developed to avoid robotic chaotic behaviors with new architectures to improve systems maintainability and traceability [9,10]. Within the field of direct image-based visual servoing systems it is worth mentioning works such as [11] that considers a FPGA (Field Programmable Gate Arrays) implementation, or [12] that extends the previous direct visual servoing for the optimal control of robot manipulators.

Within the field of indirect visual servoing, several approaches can be found for the guidance of mobile robots or mobile manipulators. In [13], a homography-based 2D approach for visual control of a mobile manipulator that does not need any measure of the 3D structure of the observed target has been proposed. A task-space sensing and control system designed to control the end effector motion of a mobile manipulator in the presence of dynamic and unknown base motion is proposed in [14]. In [15], an image-based approach with redundancy resolution is proposed for the guidance of a mobile manipulator. An indirect position-based visual control approach is described in [16] for the guidance of a mobile manipulator. In [17], a control scheme that uses image-based visual servoing was developed along with a functional mobile manipulator platform, where a hybrid camera configuration composed of monocular and stereo vision cameras was integrated into the system.

In contrast with previous approaches, the one proposed in this paper employs a direct visual servoing system for guiding mobile manipulators. In this case, the controller directly generates the torques, forces, and moments to be applied to the manipulator and base platform taking into account the robot dynamics. In the previous indirect approaches, the controller assumes that the guided device is a perfect positioning system, and therefore its dynamics is not considered. However, the use of a direct visual servo approach allows for the manipulator and base dynamics to be taken into account. The result is a faster and more accurate control that reacts more quickly to abrupt changes in the image trajectories. Additionally, an optimal control approach is used for the definition of the proposed controllers. This control approach considers the optimization of the motor signals sent to the mobile manipulator during the tracking of image trajectories. Using this optimal approach, different visual controllers with different dynamic properties can be derived.

The paper is divided into the following sections. First, Section 2 describes the robotic system and main coordinate frames. The system kinematics is described in Section 3 and the dynamics in Section 4. Section 5 describes the optimal controller and the visual servoing of the mobile manipulator. Finally, in Section 6, the main results obtained in the application of the proposed optimal controller are described.

2. System Architecture and Nomenclature

Figure 1 represents the main components of the mobile manipulator (TIAGo robot from PAL Robotics). The robot is composed of a mobile differential platform with a 7-DOF (degrees-of-freedom) arm, a RGB-D camera which is not employed in this paper, and an additional eye-in-hand camera. Frame $\mathcal{F}_p$ is attached to the mobile robot platform and $\mathcal{F}_m$ is the manipulator arm coordinate system attached to the base of the robot arm. The origin of $\mathcal{F}_m$ is translated with respect to the frame $\mathcal{F}_p$ a vector (x_{pm}, y_{pm}, z_{pm}). The world coordinate frame is called $\mathcal{F}_o$. The end effector frame $\mathcal{F}_e$ is attached to the manipulator end effector, and frame $\mathcal{F}_c$ is the eye-in-hand camera frame. In this paper we consider an eye-in-hand camera system where $\mathcal{F}_c = \mathcal{F}_e$. The orientation of the mobile platform is considered as α, i.e., the angle between de X-axis of the frame $\mathcal{F}_p$ and the X-axis of the frame $\mathcal{F}_o$ and the origin of $\mathcal{F}_p$ is at the position (x_{op}, y_{op}, z_{op}) with respect to $\mathcal{F}_o$. From the manipulator kinematics we can obtain the position of $\mathcal{F}_e$ with respect to $\mathcal{F}_m$ as (x_{me}, y_{me}, z_{me}) and $(\phi_{xme}, \phi_{yme}, \phi_{zme})$, respectively.

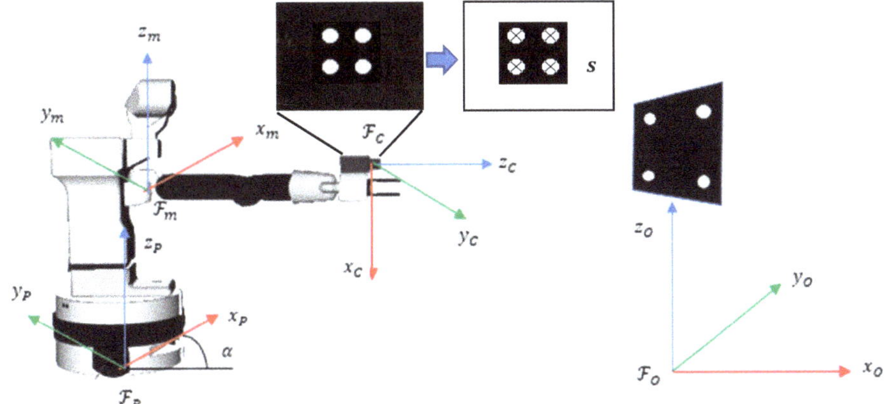

Figure 1. System components.

Regarding the visual system, an eye-in-hand camera observes a set of four visual features located on the surface of the tracked object. As Figure 1 shows, a black pattern with four white points is located in the tracked object. Therefore, the computer vision extracts four visual feature points, $s = \left[f_1 f_2 f_3 f_{4,}\right]^T$ that correspond to the center of these white points. Figure 2 shows the different modules that compose the computer vision system. The first step preprocesses the image in order to obtain a single format of the image no matter what camera is employed (e.g., image conversion from RGB to gray scale, image resizing, noise filtering, etc.). In the second stage, the image is binarized through a thresholding process. The result is a binary image with all the essential data of the pattern. The third step is an erosion operation. This is a morphological operation on the image that removes the pixels located on the edge of any object in the image. This step is mainly used to reduce the noise of the image. The fourth phase is the most important stage of the vision module: blob detection. The purpose of this module is to detect the connected pixels which form an object in the image and then assign a label to them. Finally, the vision module has to provide the pixel coordinates of the center of gravity of the visual features. The fifth submodule (area and centroid computation) accomplishes this task. In [12], a detailed description of the computer vision components required to determine the set of visual features is presented.

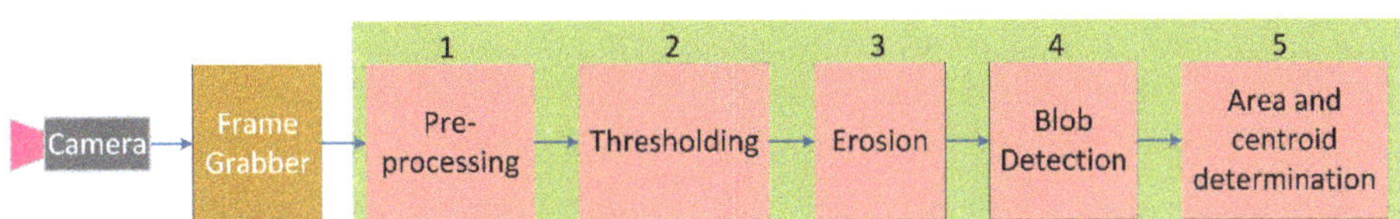

Figure 2. Computer vision components.

3. System Kinematics

The main components and coordinate frames of the robotic system are presented in Section 2. This section describes the kinematic formulation required for the definition of the proposed controllers.

$q_m^T \in \Re^{m}$ represent the generalized joint coordinates of the manipulator (in our case, m = 7) and $q_p^T \in \Re^{np}$ are the generalized coordinates of the platform. The generalized coordinates of both manipulator and platform can be represented as $q = \begin{bmatrix} q_p^T & q_m^T \end{bmatrix}$. In a general case, the computer vision system extracts k visual feature points from the observed object $s = \begin{bmatrix} f_1 f_2, \ldots, f_k \end{bmatrix}^T \in \mathbb{R}^k$ (k = 4 in the experiments, see Figure 1). Therefore, the image-based direct visual controller must perform the guidance of the mobile manipulator to track the desired trajectory in the image space, $s_d(t)$. The visual servoing control approach allows the control of the manipulator joints and base using k visual features, using an eye-in-hand configuration, where a camera is held by the end effector. Considering an image feature point, f_i, the relationship between velocities in the camera image space, $\dot{f}_i$, and the velocity twist $\dot{r}_c = \begin{bmatrix} v_c & \omega_c \end{bmatrix}^T$ of the camera (expressed in its own frame), is given by:

$$\dot{f}_i = J_{f_i}(f_i, Z_i) \begin{bmatrix} v_c \\ \omega_c \end{bmatrix} \tag{1}$$

where Z is the depth of the image feature and the value of J_{fi} can be found in [1]. From a kinematic point of view, the motor commands are considered as a set of platform command velocities, $u_p \in \mathbb{R}^p$, and manipulator command velocities, $u_m \in \mathbb{R}^m$, therefore

$$u = \begin{bmatrix} u_p^T & u_m^T \end{bmatrix}^T \tag{2}$$

In the above equation the manipulator motor commands are the joint velocities, i.e., $\dot{q}_m = u_m$, and from the kinematic model of the mobile platform the following relation can be considered:

$$\dot{q}_p = \psi\left(q_p\right) u_p \tag{3}$$

where, usually, $u_p = \begin{bmatrix} v & \omega \end{bmatrix}^T$ is the platform linear and angular velocities and the function ψ spans the admissible velocity space at each mobile platform configuration. In this paper, an image-based approach will be employed for the guidance of the mobile manipulator, therefore, an image Jacobian, J_i, must be defined which relates the differential mapping between the motor commands u and the time derivative of the set of image features s:

$$\dot{s} = J_i u \tag{4}$$

where J_i will be defined throughout the paper as the product of the matrices J_s and J_c, therefore, $J_i = J_s J_c$. The interaction matrix $J_s \in \mathbb{R}^{2k \times 6}$ for the set of image features points, s, is the stack of k matrices J_{fi} for each feature:

$$\dot{s} = J_s(s, Z) \begin{bmatrix} v_c \\ \omega_c \end{bmatrix} = \begin{bmatrix} J_{f1}(f_1, Z_1) \\ \vdots \\ J_{fk}(f_k, Z_k) \end{bmatrix} \begin{bmatrix} v_c \\ \omega_c \end{bmatrix} \tag{5}$$

where $Z = \begin{bmatrix} Z_1 & \ldots & Z_k \end{bmatrix}^T \in \mathbb{R}^k$ is a depth vector for each image feature. The matrix J_c relates the end effector velocity with respect to the camera coordinate frame $\begin{bmatrix} v_c & \omega_c \end{bmatrix}^T$ and the motor command u:

$$\begin{bmatrix} v_c \\ \omega_c \end{bmatrix} = J_c(q) u \tag{6}$$

As previously described, the mobile manipulator is composed of a differential drive platform and a 7-DOF robot arm with an eye-in-hand camera. In order to derive the value of the matrix J_c , first the position and orientation of the camera with respect to the world coordinate $\mathcal{F}_o$ is defined as

$$r = \begin{bmatrix} t^T(q) & \phi^T(q) \end{bmatrix}^T = \begin{bmatrix} t_x & t_y & t_z & \phi_x & \phi_y & \phi_z \end{bmatrix}^T \tag{7}$$

where $t^T(q)$ represents the position and $\phi^T(q)$ is the minimal parametrization of the camera orientation by Euler angles. From the robot kinematics and the system architecture presented in Section 2, it is possible to obtain

$$t_x = (x_{me} + x_{pm})cos\alpha - (y_{me} + y_{pm})cos\alpha + x_{op} \tag{8}$$

$$t_y = (x_{me} + x_{pm})sin\alpha + (y_{me} + y_{pm})cos\alpha + y_{op} \tag{9}$$

$$t_z = z_{me} + z_{pm} \tag{10}$$

From the previous equations we can obtain the time derivatives of r, i.e., $\begin{bmatrix} v^T & \omega^T \end{bmatrix}^T$. First, the components of the translation velocity, $\dot{t}^T(q)$, can be defined as

$$\dot{t}_x = \dot{x}_{me}cos\alpha - (x_{me} + x_{pm})sin(\alpha)\omega - \dot{y}_{me}sin\alpha - (y_{me} + y_{pm})cos(\alpha)\omega + vcos\alpha \tag{11}$$

$$\dot{t}_y = \dot{x}_{me}sin\alpha + (x_{me} + x_{pm})cos(\alpha)\omega + \dot{y}_{me}cos\alpha - (y_{me} + y_{pm})sin(\alpha)\omega + vsin\alpha \tag{12}$$

$$\dot{t}_z = \dot{z}_{me} \tag{13}$$

Therefore, the motion of the manipulator end depends not only on the joint positions and velocities of the manipulator arm but also on the motion of the mobile platform and its orientation α. Additionally, the components of the angular speed, $\dot{\phi}^T(q)$ of the robot can be computed as [15]

$$\dot{\phi}_x = \dot{\phi}_{xme}cos\alpha - \dot{\phi}_{yme}sin\alpha \tag{14}$$

$$\dot{\phi}_y = \dot{\phi}_{xme}sin\alpha - \dot{\phi}_{yme}cos\alpha \tag{15}$$

$$\dot{\phi}_z = \dot{\phi}_{zme} + \omega \tag{16}$$

From Equations (11)–(16) the values of $\begin{bmatrix} v^T & \omega^T \end{bmatrix}^T = \begin{bmatrix} \dot{t}_x & \dot{t}_y & \dot{t}_z & \dot{\phi}_x & \dot{\phi}_y & \dot{\phi}_z \end{bmatrix}^T$ are obtained with respect to the motor commands. Using the rotation matrix R_c, which represents the orientation frame $\mathcal{F}_c$ with respect to $\mathcal{F}_o$, the values of $\begin{bmatrix} v^T & \omega^T \end{bmatrix}^T$ can be expressed with respect to the frame $\mathcal{F}_c$, $\begin{bmatrix} v_c^T & \omega_c^T \end{bmatrix}^T$. Appendix A describes the computation of the numerical values of matrix J_c.

Another useful kinematic relationship for the definition of the proposed controllers is the image acceleration or second time derivative of s. This relationship is obtained by the definition of J_i (Equation (4)) and differentiating w.r.t.

$$\ddot{s} = J_i\dot{u} + \dot{J}_i u \tag{17}$$

4. System Dynamics

This section describes the main dynamic equations for the robotic system. The dynamic equations of motion of the mobile manipulator can be written as

$$\begin{bmatrix} F_b \\ \tau \end{bmatrix} = \begin{bmatrix} M_{bb} & M_{bm} \\ M_{bm}^T & M_{mm} \end{bmatrix} \begin{bmatrix} \dot{u}_p \\ \ddot{q}_m \end{bmatrix} + \begin{bmatrix} c_b \\ c_m \end{bmatrix} \tag{18}$$

where $\ddot{q}_m \in \Re^{m}$ is the set of joint accelerations of the robot manipulator, $\dot{u}_p = \begin{bmatrix} \dot{v} & \dot{\omega} \end{bmatrix}^T$ is the platform linear and angular accelerations, $M_{bb} \in \Re^{p\times p}$ is the inertia matrix of the platform, $M_{bm} \in \Re^{p\times m}$ is the coupled inertia matrix of the base and the manipulator, and $M_{mm} \in \Re^{m\times m}$ is the inertia matrix of the manipulator; c_b and c_m are a velocity/displacement-dependent, nonlinear terms for the base and manipulator, respectively, F_b is the force and moment exerted on the base, and τ is the applied joint torque on the robot manipulator. From a dynamic point of view, the motion of the mobile manipulator is governed by the applied torques on the manipulator joints and by the force and moment exerted on the base. Hence, Equation (18) can be written in the following compact form.

$$\widetilde{\tau} = \widetilde{M}\dot{u} + \widetilde{C} \tag{19}$$

where $\widetilde{\tau} = \begin{bmatrix} F_b \\ \tau \end{bmatrix}$, $\widetilde{M} = \begin{bmatrix} M_{bb} & M_{bm} \\ M_{bm}^{\mathrm{T}} & M_{mm} \end{bmatrix}$, $\dot{u} = \begin{bmatrix} \dot{u}_p \\ \ddot{q}_m \end{bmatrix}$, and $\widetilde{C} = \begin{bmatrix} c_b \\ c_m \end{bmatrix}$.

5. Optimal Control of the Mobile Manipulator

5.1. Optimal Control Definition

In complex high dimensional systems, several tasks can be accomplished simultaneously taking advantage of the robot redundancy. This paper considers a system with μ constrains that represents the task for the mobile manipulator to be executed. These constraints can be represented as

$$A\ (t)\dot{u} = b(t) \tag{20}$$

where $A\ (t) \in \Re^{\mu\times(p+m)}$ and $b\ (t) \in \Re^{\mu\times 1}$. An advantage of this task formulation is that nonholonomic constraints can be treated in the same general way. In order to reduce the energy required for performing the visual servoing task, the proposed optimal controller is designed to minimize the control torque for the mobile manipulator, considering the following cost function.

$$\Omega(t) = \widetilde{\tau}^{\mathrm{T}}\mathbf{W}(t)\widetilde{\tau} \tag{21}$$

where $\mathbf{W}(t)$ is a time-dependent weight matrix.

Let us assume that the robot model presented in the previous section and a constraint formulation are given. In this case, the optimal controller has to guarantee that the task is perfectly achieved, i.e., that $A\ (t)\dot{u} = b(t)$ holds at all times. Additionally, the minimization of the control command with respect to some given metric, i.e., $\Omega(t) = \widetilde{\tau}^{\mathrm{T}}\mathbf{W}(t)\widetilde{\tau}$, is intended at each instant of time. The solution to this pointwise optimal control problem [18] can be derived from a generalization of Gauss' principle, as originally suggested in [19]. It is also a generalization of the propositions in [20,21], where the following control action is obtained considering the general expression for the dynamic model expressed in Equation (19) (for the sake of clarity the time dependences are not indicated):

$$\widetilde{\tau} = \mathbf{W}^{-1/2}\left(A\widetilde{M}^{-1}\mathbf{W}^{-1/2}\right)^{+}\cdot\left(b + A\widetilde{M}^{-1}\widetilde{C}\right) \tag{22}$$

where the symbol + denotes the pseudo inverse of a general matrix. As it can be seen in Equation (22), the matrix **W** is an important factor in the control law that can be used to determine how the control efforts are distributed over the joints and the robot base.

5.2. Optimal Direct Visual Servoing

Once the optimal control framework is presented in Section 5.1, this section describes its application for the optimal control of the mobile platform. First, we define $\ddot{s}_r$ as the reference image accelerations that will be employed by the optimal controller. The value of these image accelerations can be obtained from Equation (17). This last expression can be expressed into the form of Equation (20) as

$$J_i\dot{u} = \ddot{s}_r - \dot{J}_i u \tag{23}$$

This way, the visual servoing task is defined by the following relationships.

$$A = J_i b = \ddot{s}_r - \dot{J}_i u \tag{24}$$

Therefore, with the given definition of A and b, the optimal control will minimize the motor commands while performing the tracking in the image space. The final control law can be obtained replacing these variables into the function that minimizes the motor signals described by Equation (22):

$$\tilde{\tau} = \mathbf{W}^{-1/2}\left(J_i\tilde{M}^{-1}\mathbf{W}^{-1/2}\right)^{+}\cdot\left(\ddot{s}_r - \dot{J}_i u + J_i\tilde{M}^{-1}\tilde{C}\right) \tag{25}$$

As it can be seen, the visual controller represented by (25) implicitly depends on the weighting matrix **W**, and different values of this matrix can be used to coordinate the motion of the base and of the manipulator. The diagram of the considered control loop is depicted in Figure 3.

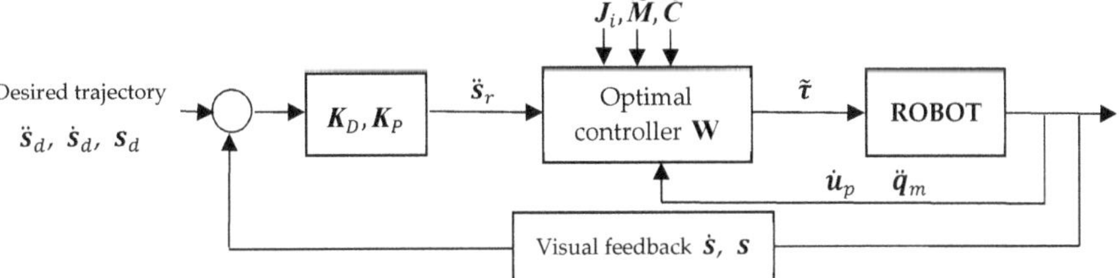

Figure 3. Diagram of the control scheme for the mobile manipulator robot.

Considering $\mathbf{W} = \tilde{M}^{-2}$ and replacing this value in the control framework expressed in Equation (25), the result yields

$$\tilde{\tau} = \tilde{M}J_i^{+}\cdot\left(\ddot{s}_r - \dot{J}_i u + J_i\tilde{M}^{-1}\tilde{C}\right) \tag{26}$$

This controller represents a direct visual servo control using inversion of the dynamic model. Additionally, new controllers can be obtained using different values of **W**. For example, an important value for **W**, due to its physical interpretation, is $\mathbf{W} = \tilde{M}^{-1}$, since it is consistent with the principle of d'Alembert:

$$\tilde{\tau} = \tilde{M}^{1/2}\left(J_i\tilde{M}^{-1}\tilde{M}^{1/2}\right)^{+}\cdot\left(\ddot{s}_r - \dot{J}_i u + J_i\tilde{M}^{-1}\tilde{C}\right) \tag{27}$$

Now, the definition of the reference control, $\ddot{s}_r$, is described considering the eye-in-hand camera system that extracts a set of k image feature points. The task description as a constraint is given by the following equation in the image space.

$$(\ddot{s}_d - \ddot{s}) + K_D(\dot{s}_d - \dot{s}) + K_P(s_d - s) = 0 \tag{28}$$

where $\ddot{s}_d$, $\dot{s}_d$, and s_d are the desired image space accelerations, velocities, and positions, respectively. K_P and K_D are proportional and derivative gain matrices, respectively. This equation can be expressed in regards to image error in the following way.

$$\ddot{s}_d + K_D\dot{e}_s + K_P e_s = \ddot{s}_r \tag{29}$$

where e_s and $\dot{e}_s$ are the image error and the time derivative of the image error, respectively. As stated, the variable $\ddot{s}_r$ denotes the reference image accelerations of the proposed image space based controller. Substituting this variable into the dynamic visual servo controller in Equation (25), the control law is set by the following relationship.

$$\tilde{\tau} = \mathbf{W}^{-1/2}\left(J_i\tilde{M}^{-1}\mathbf{W}^{-1/2}\right)^{+}\cdot\left(\ddot{s}_d + K_D\dot{e}_s + K_P e_s - \dot{J}_i u + J_i\tilde{M}^{-1}\tilde{C}\right) \tag{30}$$

In order to demonstrate the asymptotic tracking of the control law, some operations must be done. First, the closed-loop behavior is computed from Equation (19) as

$$\tilde{M}\dot{u} + \tilde{C} = \mathbf{W}^{-1/2}\left(J_i\tilde{M}^{-1}\mathbf{W}^{-1/2}\right)^{+}\cdot\left(\ddot{s}_d + K_D\dot{e}_s + K_P e_s - \dot{J}_i u + J_i\tilde{M}^{-1}\tilde{C}\right) \tag{31}$$

Equation (31) can be simplified by premultiplying its left and right sides by $\left(J_i\tilde{M}^{-1}\mathbf{W}^{-1/2}\right)\mathbf{W}^{1/2}$:

$$J_i\dot{u} = \ddot{s}_d + K_D\dot{e}_s + K_P e_s - \dot{J}_i u \tag{32}$$

Finally, using the relationship expressed in (23), it can be concluded that

$$\ddot{e}_s = -K_D\dot{e}_s - K_P e_s \tag{33}$$

Therefore, when J_i is full-rank, an asymptotic tracking is achieved by the visual servo controller expressed in Equation (30) for the tracking of an image trajectory.

6. Results

This section presents the system behavior during the tracking of different trajectories. The robot is guided by an eye-in-hand camera system. The parameters of a Gigabit Ethernet TM6740GEV camera is considered, which acquires 200 images every second with a resolution of 1280 × 1024 pixels. The eye-in-hand camera extracts four visual features from the four corners of the target.

Three different experiments are presented in this section. In these experiments the value of $\mathbf{W} = M^{-1}$ is considered. Figure 4a,b represents the initial configuration employed in all of the experiments. Figure 4a represents the target and the mobile manipulator with the eye-in-hand camera system. The initial position of the visual features is represented in Figure 4b. The desired trajectory employed in the first experiment is a linear trajectory that represents a lateral translation of 200 px. in x and y directions in the image from the initial position of the visual features. As it can be seen in Figure 4d, the image-based controller correctly carries out the tracking and a straight line is obtained in the image space. Figure 4c represents the final pose of the mobile manipulator. The second experiment, represented in Figure 4e,f, evaluates the tracking when a displacement in depth must be performed. As it can be seen in Figure 4e, a displacement in depth is carried out and, once the experiment ends, the mobile manipulator is closer to the target. Figure 4f represents the obtained image trajectory. This trajectory corresponds to the desired trajectory specified as a linear increase in depth from the initial image configuration. Finally, the experiment represented in Figure 4g,h requires a change in orientation of the eye-in-hand camera with respect to the observed target. To do this, the desired image trajectories for the visual features are represented by a straight line between the initial and the final ones defined as $s^* = [f_{1d} = (503, 584), f_{2d} = (612, 600), f_{3d} = (523, 472), f_{4d} = (630, 477)]^T$ px. As in the previous experiments, Figure 4g represents the final robot pose once the experiment ends and Figure 4h represents the image trajectory. As the experiments show, the proposed controller allows the concurrent guidance of both the manipulator and robotic platform during the tracking of the desired image trajectory.

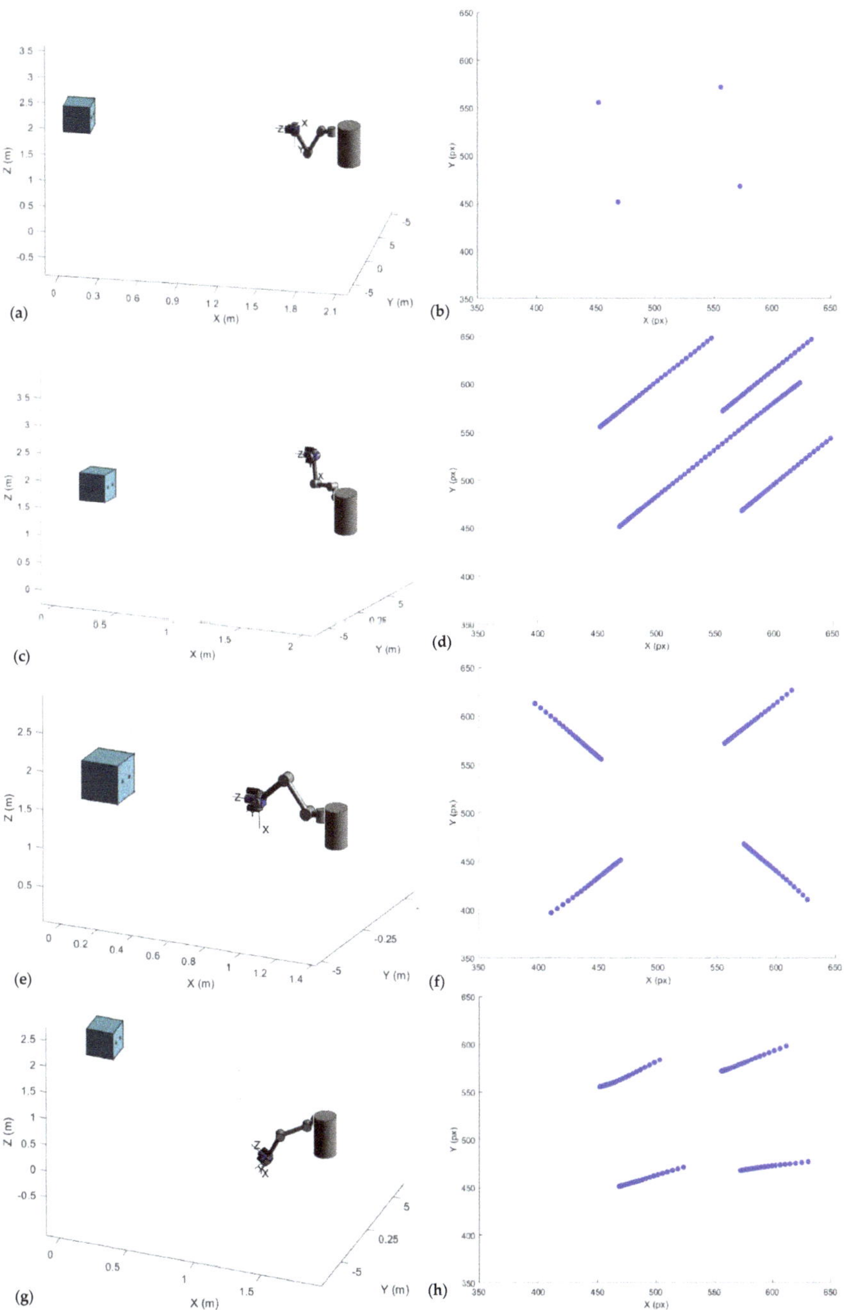

Figure 4. Robot pose and image trajectories in the experiments. (**a**) 3D initial pose of the mobile manipulator and target. (**b**) Initial visual features extracted by the eye-in-hand camera. First experiment: (**c**) 3D final pose. (**d**) Image trajectory. Second experiment: (**e**) 3D final pose. (**f**) Image trajectory. Third experiment: (**g**) 3D Final pose. (**h**) Image trajectory.

Figure 5 represents the joint torques applied to the manipulator (m = 1 ... 7) and force and moments applied to the robot platform during the first and second experiments. As it can be seen in Figure 5a (first experiment), the torques, force, and moment remain low, and a smooth behavior is obtained during the tracking. For the second experiment, Figure 5b represent the joint torques applied to the manipulator and force and moment applied to the base platform during the robot motion in depth. It is worth noting that the reduction of the distance between the two bodies (target and robot) is obtained by modifying both the base platform pose and the robot manipulator joint configuration. Finally, as in the previous experiments, Figure 6 represents the joint torques applied to the manipulator and force and moment applied to the base platform during the third tracking experiment. The desired relative orientation is achieved by controlling both the orientation of the base platform and the robot manipulator joint configuration.

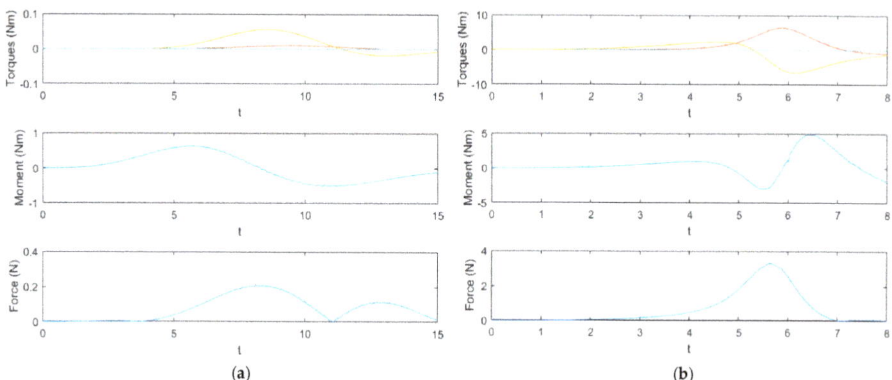

Figure 5. (**a**) Experiment 1: Joint torque and moment and force exerted on the base platform. (**b**) Experiment 2: Joint torque, moment, and force exerted on the base platform.

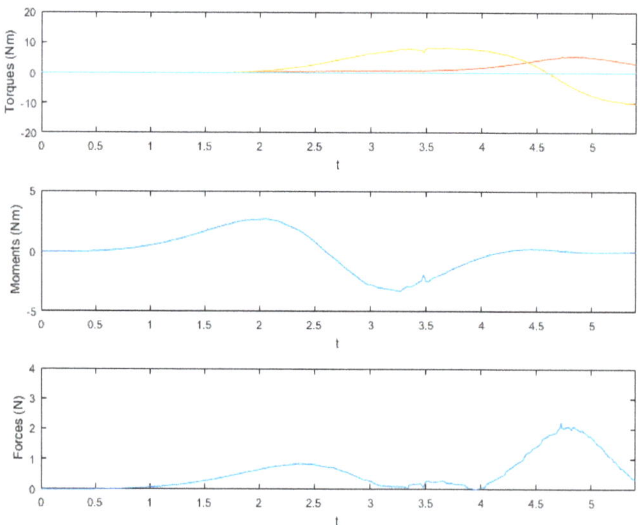

Figure 6. Experiment 3: Joint torque, moment, and force exerted on the base platform.

One of the main advantages of the proposed controller is the possibility to define new direct image-based controllers by modifying the value of the matrix **W** of the optimal controller (see Equation (25)). A different controller with different dynamic behavior during the tracking of image trajectories can be obtained by modifying this matrix. In order to evaluate the proposed framework, in Table 1 we represent the obtained image error during the tracking of a repetitive image trajectory defined by the following equation.

$$s_d = \begin{bmatrix} f_{xd} \\ f_{yd} \end{bmatrix} = \begin{bmatrix} 320 + 166cos(\omega t + \pi/4) \\ 265 + 160sin(\omega t + \pi/4) \end{bmatrix} \quad (34)$$

Table 1. Mean image error in pixels (mm) during the tracking of the image trajectory.

W	$\omega = 0.5$rad/s	$\omega = 1$rad/s	$\omega = 2$rad/s
$\mathbf{W} = \widetilde{M}^{-2}$	2.14 px (0.59 mm)	2.21 px (0.61 mm)	3.84 px (1.04 mm)
$\mathbf{W} = \widetilde{M}^{-1}$	2.12 px (0.58 mm)	2.95 px (0.8 mm)	2.39 px (0.66 mm)
$\mathbf{W} = I$	3.81 px (1.03 mm)	3.22 px (0.87 mm)	4.4 px (1.2 mm)
Previous	2.25 px (0.62 mm)	3.2 px 0.85 mm)	4.12 px (1.09 mm)

Table 1 represents the mean error in pixels and mm during the tracking of the image trajectory considering different tracking velocities. As it can be seen in this table, different tracking errors are obtained and the best behavior at high velocities is obtained considering $\mathbf{W} = \widetilde{M}^{-1}$. Additionally, the performance of the controllers is compared with respect to previous controllers only based on inverse dynamics [22] (indicated as "previous" in Table 1).

7. Conclusions

This paper presents a direct image-based visual servoing system for the guidance of mobile manipulators. This approach is based on an optimal framework that considers the optimization of the motor signals sent to the mobile manipulator during the tracking of image trajectories. The stability of the controller has been proved analytically, and different dynamical behavior can be obtained by using different values of the matrix **W**.

As a result, an adaptive and flexible tool has been obtained, allowing for concurrent control of both the base platform and manipulator during the tracking of trajectories. With the given definition of A and $\boldsymbol{b}$, the optimal control can minimize the motor commands while performing the tracking of trajectories in the image space. The viability of the controller has been tested in a variety of test case experiments, including approaches to an object, as well as different positioning tasks that require modifying the position and/or the orientation of the mobile manipulator. In all such cases, the experiments and trajectories have been accomplished.

Author Contributions: Funding acquisition, C.A.J.; Methodology, J.L.R.; Software, Á.B.; Supervision, J.P.; Validation, G.J.G.

Funding: This research was supported by the Valencia Regional Government through project GV/2018/050.

Conflicts of Interest: The authors declare no conflict of interest.

Appendix A

As described throughout the paper, the motor commands are $\boldsymbol{u} = \begin{bmatrix} \boldsymbol{u}_p^T & \boldsymbol{u}_m^T \end{bmatrix}^T = \begin{bmatrix} v & \omega & \dot{q}_m^T \end{bmatrix}^T$. First, the robot Jacobian is used to convert the joint velocities to Cartesian velocities:

$$\begin{bmatrix} v \\ \omega \\ \dot{x}_{me} \\ \dot{y}_{me} \\ \dot{z}_{me} \\ \dot{\phi}_{xme} \\ \dot{\phi}_{yme} \\ \dot{\phi}_{zme} \end{bmatrix} = \begin{bmatrix} \boldsymbol{I}_{2\times 2} & \boldsymbol{0}_{2\times 6} \\ \boldsymbol{0}_{6\times 2} & \boldsymbol{J}_{6\times 6} \end{bmatrix} \boldsymbol{u}. \tag{A1}$$

The relation between the previous Cartesian velocities and $\begin{bmatrix} \boldsymbol{v}^T & \boldsymbol{\omega}^T \end{bmatrix}^T$ can be obtained by using Equations (11)–(16):

$$\begin{bmatrix} \dot{t}_x \\ \dot{t}_y \\ \dot{t}_z \\ \dot{\phi}_x \\ \dot{\phi}_y \\ \dot{\phi}_z \end{bmatrix} = \begin{bmatrix} J_{c13\times 8} \\ J_{c23\times 8} \end{bmatrix} \begin{bmatrix} v \\ \omega \\ \dot{x}_{me} \\ \dot{y}_{me} \\ \dot{z}_{me} \\ \dot{\phi}_{xme} \\ \dot{\phi}_{yme} \\ \dot{\phi}_{zme} \end{bmatrix} \tag{A2}$$

Where,

$$J_{c1} = \begin{bmatrix} cos\alpha & -x_{me}sin\alpha - x_{pm}sin\alpha & cos\alpha & -sin\alpha & 0 & 0 & 0 & 0 \\ sin\alpha & x_{me}cos\alpha + x_{pm}cos\alpha & sin\alpha & cos\alpha & 0 & 0 & 0 & 0 \\ 0 & 0 & 0 & 0 & 1 & 0 & 0 & 0 \end{bmatrix}$$
$$J_{c2} = \begin{bmatrix} 0 & 0 & 0 & 0 & 0 & cos\alpha & -sin\alpha & 0 \\ 0 & 0 & 0 & 0 & 0 & sin\alpha & -cos\alpha & 0 \\ 0 & 1 & 0 & 0 & 0 & 0 & 0 & 1 \end{bmatrix} \tag{A3}$$

Finally, by using the rotation matrix $\boldsymbol{R}_c$ which represents the orientation frame $\mathcal{F}_c$ with respect to the fame $\mathcal{F}_o$, the values of $\begin{bmatrix} \boldsymbol{v}^T & \boldsymbol{\omega}^T \end{bmatrix}^T$ can be expressed with respect to the frame $\mathcal{F}_c$, $\begin{bmatrix} \boldsymbol{v}_c^T & \boldsymbol{\omega}_c^T \end{bmatrix}^T$.

$$\begin{bmatrix} \boldsymbol{v}_c \\ \boldsymbol{\omega}_c \end{bmatrix} = \begin{bmatrix} R_c^T & \boldsymbol{0} \\ \boldsymbol{0} & R_c^T \end{bmatrix} \begin{bmatrix} \boldsymbol{v} \\ \boldsymbol{\omega} \end{bmatrix} \tag{A4}$$

Therefore, the value of J_c is equal to

$$J_c = \begin{bmatrix} R_c^T & \boldsymbol{0} \\ \boldsymbol{0} & R_c^T \end{bmatrix} \begin{bmatrix} J_{c13\times 8} \\ J_{c23\times 8} \end{bmatrix} \begin{bmatrix} \boldsymbol{I}_{2\times 2} & \boldsymbol{0}_{2\times 2} \\ \boldsymbol{0}_{6\times 2} & \boldsymbol{J}_{6\times 6} \end{bmatrix} \tag{A5}$$

References

1. Chaumette, F.; Hutchinson, S. Visual servo control. I. Basic approaches. *IEEE Robot. Autom. Mag.* **2006**, *13*, 82–90. [CrossRef]
2. Allen, P.K.; Yoshimi, B.; Timcenko, A. Real-time visual servoing. In Proceedings of the 1991 IEEE International Conference on Robotics and Automation, Sacramento, CA, USA, 9–11 April 1991; pp. 851–856.
3. Huang, P.; Zhang, F.; Cai, J.; Wang, D.; Meng, Z.; Guo, J. Dexterous Tethered Space Robot: Design, Measurement, Control, and Experiment. *IEEE Trans. Aerosp. Electron. Syst.* **2017**, *53*, 1452–1468. [CrossRef]

4. Park, D.; Kwon, J.; Ha, I. Novel Position-Based Visual Servoing Approach to Robust Global Stability Under Field-of-View Constraint. *IEEE Trans. Ind. Electron.* **2012**, *59*, 4735–4752. [CrossRef]
5. Chaumette, F.; Hutchinson, S. Visual control. II. Advanced approaches. *IEEE Robot. Autom. Mag.* **2007**, *14*, 109–118. [CrossRef]
6. Bateux, Q.; Marchand, E.; Leitner, J.; Chaumette, F.; Corke, P. Training Deep Neural Networks for Visual Servoing. In Proceedings of the 2018 IEEE International Conference on Robotics and Automation (ICRA), Brisbane, Australian, 21–25 May 2018; pp. 1–8.
7. Levine, S.; Pastor, P.; Krizhevsky, A.; Ibarz, J.; Quillen, D. Learning hand-eye coordination for robotic grasping with deep learning and large-scale data collection. *Int. J. Robot. Res.* **2018**, *37*, 421–436. [CrossRef]
8. Agravante, D.J.; Claudio, G.; Spindler, F.; Chaumette, F. Visual Servoing in an Optimization Framework for the Whole-Body Control of Humanoid Robots. *IEEE Robot. Autom. Lett.* **2017**, *2*, 608–615. [CrossRef]
9. Pomares, J.; Jara, C.A.; Perez, J.; Torres, F. Direct visual servoing framework based on optimal control for redundant joint structures. *Int. J. Precis. Eng. Manuf.* **2015**, *16*, 267–274. [CrossRef]
10. Pomares, J.; Perea, I.; Torres, F. Dynamic Visual Servoing With Chaos Control for Redundant Robots. *IEEE/ASME Trans. Mechatron.* **2014**, *19*, 423–431. [CrossRef]
11. Alabdo, A.; Pérez, J.; Garcia, G.J.; Pomares, J.; Torres, F. FPGA-based architecture for direct visual control robotic systems. *Mechatronics* **2016**, *39*, 204–216. [CrossRef]
12. Pérez, J.; Alabdo, A.; Pomares, J.; García, G.J.; Torres, F. FPGA-based visual control system using dynamic perceptibility. *Robot. Comput. Integr. Manuf.* **2016**, *41*, 13–22. [CrossRef]
13. Benhimane, S.; Malis, E. Homography-based 2D Visual Tracking and Servoing. *Int. J. Robot. Res.* **2007**, *26*, 661–676. [CrossRef]
14. Sandy, T.; Buchli, J. Dynamically decoupling base and end-effector motion for mobile manipulation using visual-inertial sensing. In Proceedings of the 2017 IEEE/RSJ International Conference on Intelligent Robots and Systems (IROS), Vancouver, BC, Canada, 24–28 September 2017; pp. 6299–6306.
15. Luca, A.D.; Oriolo, G.; Giordano, P.R. Image-based visual servoing schemes for nonholonomic mobile manipulators. *Robotica* **2007**, *25*, 131–145.
16. Lang, H.; Khan, M.T.; Tan, K.K. Application of visual servo control in autonomous mobile rescue robots. *Int. J. Comput. Commun. Control* **2016**, *11*, 685–696. [CrossRef]
17. Wang, Y.; Zhang, G.; Lang, H. A modified image-visual servo controller with hybrid camera configuration for robust robotic grasping. *Robot. Auton. Syst.* **2014**, *62*, 1398–1407. [CrossRef]
18. Spong, M.; Thorp, J.; Kleinwaks, J. The control of robot manipulators with bounded input. *IEEE Trans. Autom. Control* **1986**, *31*, 483–490. [CrossRef]
19. Udwadia, F.E. A new perspective on tracking control of nonlinear structural and mechanical systems. *Proc. R. Soc. Lond. Ser. A* **2003**, 1783–1800. [CrossRef]
20. Udwadia, F.E.; Kalaba, R.E. *Analytical Dynamics: A New Approach*; Cambridge University Press: Cambridge, UK, 1996.
21. Bruyninckx, H.; Khatib, O. Gauss' principle and the dynamics of redundant and constrained manipulators. In Proceedings of the 2000 IEEE International Conference on Robotics & Automation, San Francisco, CA, USA, 24–28 April 2000; pp. 2563–2569.
22. Pomares, J.; Corrales, J.A.; García, G.J.; Torres, F. Direct visual Servoing to track trajectories in human-robot cooperation. *Int. J. Adv. Robot. Syst.* **2011**, *8*, 44. [CrossRef]

Article

Picking Robot Visual Servo Control Based on Modified Fuzzy Neural Network Sliding Mode Algorithms

Wei Chen [1,2,*], Tongqing Xu [1], Junjie Liu [1], Mo Wang [2] and Dean Zhao [3]

1 School of Electronics and Information, Jiangsu University of Science and Technology, Zhenjiang 212003, China; 179030038@stu.just.edu.cn (T.X.); junjiel_mtr@163.com (J.L.)
2 Microrobotics Lab, The University of Sheffield, Sheffield S13JD, UK; mwang56@sheffield.ac.uk
3 School of Electrical and Information Engineering, Jiangsu University, Zhenjiang 212003, China; dazhao@ujs.edu.cn
* Correspondence: cwchenwei@aliyun.com or W.Chen2@sheffield.ac.uk

Received: 23 April 2019; Accepted: 24 May 2019; Published: 29 May 2019

Abstract: Through an analysis of the kinematics and dynamics relations between the target positioning of manipulator joint angles of an apple-picking robot, the sliding-mode control (SMC) method is introduced into robot servo control according to the characteristics of servo control. However, the biggest problem of the sliding-mode variable structure control is chattering, and the speed, inertia, acceleration, switching surface, and other factors are also considered when approaching the sliding die surface. Meanwhile, neural network has the characteristics of approaching non-linear function and not depending on the mechanism model of the system. Therefore, the fuzzy neural network control algorithm can effectively solve the chattering problem caused by the variable structure of the sliding mode and improve the dynamic and static performances of the control system. The comparison experiment is carried out through the application of the PID algorithm, the sliding mode control algorithm, and the improved fuzzy neural network sliding mode control algorithm on the picking robot system in the laboratory environment. The result verified that the intelligent algorithm can reduce the complexity of parameter adjustments and improve the control accuracy to a certain extent.

Keywords: visual servoing; fuzzy neural network; sliding mode control; picking robot

---

1. Introduction

With the development of science and technology, robot technology has acquired unprecedented achievements. In agriculture, the development of fruit-picking robots has become a hotpot project due to the rising labor costs and the huge output of apples and citrus. The apple-picking robot is an integrated system that integrates environmental perception, dynamic decision-making, and behavior control; motion control is therein the most basic and important link. At present, there are two main kinds of visual servo control for picking robots: one is image-based visual servo control, and the other is a position-based visual servo control system [1]. The picking robot manipulator is a typical non-linear system, because it is composed of multiple joints and has time-varying uncertainty during the picking process. This indicates that traditional control methods would not meet the control requirements. Image feedback-based visual servo control is a real-time feedback control for the manipulator according to the visual acquisition image, and it also shows strong self-adaptability to the change of external environment and system parameters so that it reduces the complexity of manipulator in the control process.

At present, most visual servo control systems based on image feedback adopt traditional control schemes such as adaptive control [2], PID control [3], and sliding model control [4], etc., or intelligent control algorithms [5] such as neural network and fuzzy theory [6], etc. There are some defects in adopting a single control method. Adaptive control is a strategy to deal with structural uncertainties, which provides good performance to a certain extent [7]. PID control has better control performance for the position and joint angle parameters of the manipulator end-effector [8]. Sliding-mode control (SMC) has excellent robustness and good anti-jamming ability to the uncertainties of the system, especially in the control of non-linear systems [9]. Neural network (NN) has the ability to learn and approximate non-linear functions [10]. It is often used to compensate unstructured uncertainties in robot controller modeling. Fuzzy control [11] imitates human reasoning and decision-making processes on the basis of not depending on the precise mathematical model of the controlled object. Fuzzy logic is adopted in robot controllers to achieve good control performance under uncertain conditions. In the reference [12], a robust adaptive control method based on the dynamic structure of the fuzzy wavelet neural network is proposed, which is used to track the trajectory of the robot with uncertainties and disturbances by the adaptive sliding-mode control. Mehta and Ton et al. introduced a continuous terminal sliding mode visual servo controller to adjust the camera to the desired position, but this has many requirements for the camera, and requires the correction of the camera position [13]. In the field of robotic citrus picking, there is a non-linear robust image visual servo controller that can adjust the end effector to the fruit position under the condition of unknown fruit movement, but the speed of the picking process is slow, and is not conducive to agricultural application [14]. The fuzzy neural network (FNN) learning algorithm [15,16] is a superior system identification method that utilizes non-linear control. There is no need to get the prior knowledge about the uncertainty and a sufficient amount of observed data. Efe and Devesh et al. used the novel particle swarm optimization (PSO)–SMC hybrid learning algorithm to train the FNN effectively, but instead of using the visual servo control of picking robots, they realized the flattening and rotation control of the four-rotor configuration for agricultural purposes [17].

A great deal of research has been devoted to these intelligent algorithm scholars, but there are few applications in the field of agricultural picking robots. In this paper, a joint visual servo control algorithm based on an improved sliding mode and kinematics and dynamics equation is presented that appears to have good robustness in a non-linear system. However, sliding-mode control always requires detailed system information and corresponding uncertainty bounds to ensure stability [18]. Even though the stability requirements are met, there will still be problems such as jitter [19]. Therefore, fuzzy and adaptive neural network control algorithms are introduced and combined with sliding-mode control to suppress the sliding-mode jitter problem. However, the traditional neural network model is complex and computationally intensive, with a high probability of over-fitting. To prevent over-fitting, we increased the dropout rate for the dropout layer [20].

According to the characteristics of apple-picking robot motion control, the image-based robot visual servo control method is combined with the improved fuzzy neural network sliding mode control algorithm to make use of the non-linear and non-system-dependent mechanism model of the fuzzy neural network. Combined with sliding mode variable structure control, the sliding mode chattering is eliminated, and the stability of the control system is improved. In this paper, the algorithm is applied for the first time in the field of agricultural fruit-picking robots, which improves the extraction speed of apple-picking robots.

The research aim of this paper is mainly to verify the visual servo control of the sliding-mode control algorithm based on the improved fuzzy neural network through the apple-picking robot. The article is divided into three parts, Firstly, we establish the kinematics and dynamics equations of the picking robot. Secondly, visual positioning is introduced to calculate the position of the target point, and an image-based visual control algorithm is used. Finally, the sliding-mode control algorithm and the improved fuzzy neural network control algorithm are combined to carry out simulation analysis, and the algorithm is verified on the picking robot. It is concluded that the improved fuzzy neural network sliding mode control algorithm can improve the efficiency of the robot arm servo control and has higher stability.

2. Picking Robot Structure and Model Establishment

2.1. Manipulator Structure Design

Since the apple trees are relatively high and the fruits are disperse, the higher degree of freedom of the picking robot manipulator brings not only the kinematics algorithm complexity, but also the waste of joint freedom degree, which is disadvantageous to the practicality of agriculture. This reduces the requirements of the manipulator and minimizes the usage of freedom degrees. So, this paper introduces the self-designed five degrees of freedom (DOF) manipulator, which is shown in Figure 1. Figure 2 is a schematic view of the picking robot manipulator. The manipulator gripping device adopts a curved design in order to reduce the damage of the apple during the process of picking up the apple. Figure 3 shows a wheeled picking robot.

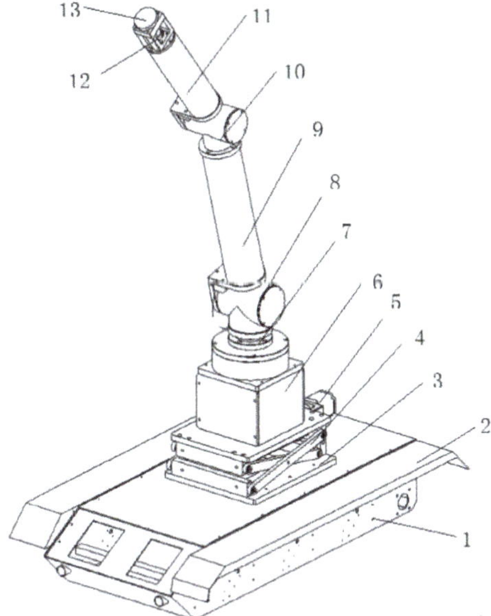

Figure 1. Manipulator structural drawing of picking robot (1: picking robot platform; 2: platform; 3: robot base; 4: lift platform; 5: servo motor; 6: robot lumbar; 7: lumbar rotating shaft; 8: main arm joint; 9: main arm; 10: small arm joint; 11: small arm; 12: rotating shaft; 13: grasp hand installation seat).

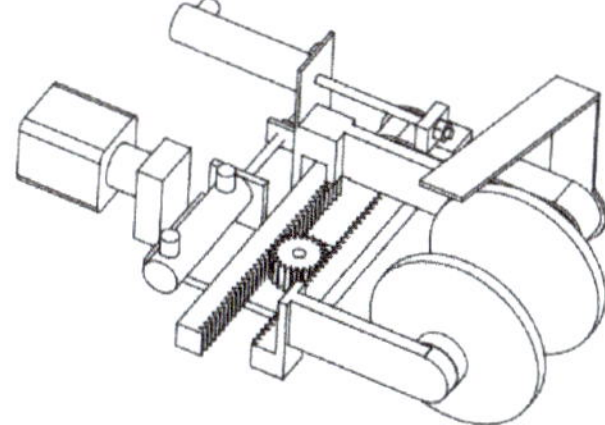

Figure 2. Schematic diagram of picking robot manipulator.

Figure 3. Picking robot physical map.

Fruit-picking robots are mainly composed of two parts. The first part is a chassis platform, and the second part is a manipulator. Therein, the five DOF manipulator is a PRRRP joint structure. The first degree of freedom is the lifting platform. Its main function is to lift the whole manipulator. The second degree of freedom is the waist rotation joint of the manipulator. The third degree of freedom is the swing axis of the arm, the fourth degree of freedom is the swing axis of the arm, and the fifth degree of freedom is the rotating axis of the manipulator with the function of adjusting the grasping posture according to the picking requirement.

2.2. Kinematics Equation of Manipulator

The manipulator is a complex system, since it includes various disciplines such as kinematics and dynamics. Kinematics analysis is used to model according to the relationship between the spatial posture and the angle of each joint of the manipulator. The designed structural kinematics parameters of the manipulator are shown in Table 1. In the table, α_i is the torsion angle, a_i is the length of the connecting rod, θ_i is the joint angle, d_i is the offset of the connecting rod, $\sigma = 0$ is the rotating joint, $\sigma = 1$ is the direct acting joint, and d_i is the other constant. The use of the D–H rule to determine the coordinate position of the connecting rod is shown in Figure 4.

Table 1. Kinematic parameters of manipulators.

Joints i	$\alpha_i/(°)$	$a_i/(°)$	$\theta_i/(°)$	$d_i/(°)$	σ	Variable Range
1	0	0	90	d_1	1	d_1(0.63 to 1.3)
2	−90	0.133	θ_2	0	0	θ_2 (−180° to 180°)
3	0	1	θ_3	0	0	θ_3 (−142° to −36°)
4	−90	0	θ_4	0	0	θ_4 (−180° to 180°)
5	0	0	θ_5	0	0	θ_5 (0° to 360°)

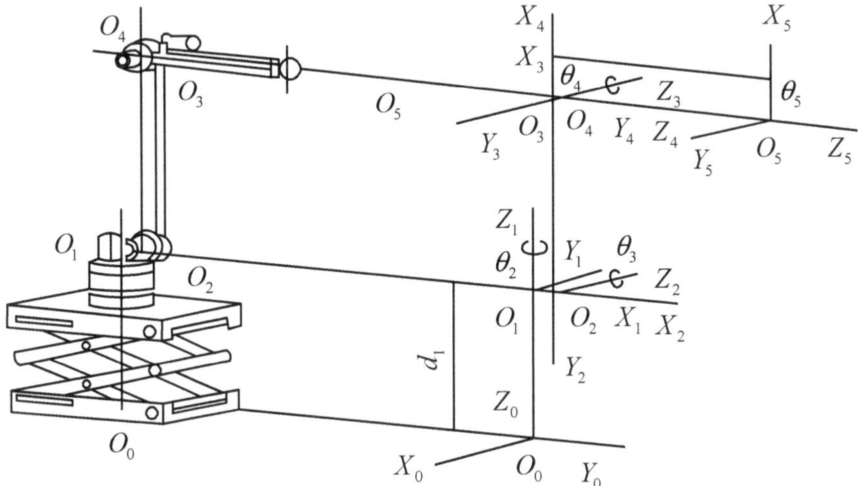

Figure 4. Manipulator D–H model.

The homogeneous coordinate transformation matrix is used in the above-mentioned establishment of connecting the rod D–H coordinate system and kinematics mode.

$$
\begin{gathered}
{}^{0}_{1}T = \begin{bmatrix} 0 & -1 & 0 & 0 \\ 1 & 0 & 0 & 0 \\ 0 & 0 & 1 & d_1 \\ 0 & 0 & 0 & 1 \end{bmatrix} \\
{}^{1}_{2}T = \begin{bmatrix} cos\theta_2 & 0 & -sin\theta_2 & 0.133\,cos\theta_2 \\ sin\theta_2 & 0 & cos\theta_2 & 0.133\,sin\theta_2 \\ 0 & -1 & 0 & 0 \\ 0 & 0 & 0 & 1 \end{bmatrix} \\
{}^{2}_{3}T = \begin{bmatrix} cos\theta_3 & -sin\theta_3 & 0 & cos\theta_3 \\ 1 & cos\theta_3 & 0 & sin\theta_3 \\ 0 & 0 & 1 & 0 \\ 0 & 0 & 0 & 1 \end{bmatrix} \\
{}^{3}_{4}T = \begin{bmatrix} cos\theta_4 & 0 & -sin\theta_4 & 0 \\ sin\theta_4 & 0 & cos\theta_4 & 0 \\ 0 & -1 & 0 & 0 \\ 0 & 0 & 0 & 1 \end{bmatrix} \\
{}^{4}_{5}T = \begin{bmatrix} 1 & 0 & 0 & 0 \\ 0 & 1 & 0 & 0 \\ 0 & 0 & 1 & d_5 \\ 0 & 0 & 0 & 1 \end{bmatrix}
\end{gathered} \tag{1}
$$

Therein, ${}^{i-1}_{i}T$ presents the homogeneous matrix of translation and rotation from the $\boldsymbol{i} - 1$ joint coordinate system to the $\boldsymbol{i}$ joint coordinate system. The total transformation matrix of the end effector relative to the robot coordinate system is:

$$
{}^{0}_{5}T = {}^{0}_{1}T{}^{1}_{2}T{}^{2}_{3}T{}^{3}_{4}T{}^{4}_{5}T = \begin{bmatrix} n_x & o_x & a_x & p_x \\ n_y & o_y & a_y & p_y \\ n_z & o_z & a_z & p_z \\ 0 & 0 & 0 & 1 \end{bmatrix} \tag{2}
$$

The end effector position equation is:

$$\begin{aligned} p_x &= d_5\, sin\theta_2 sin(\theta_3+\theta_4) - sin\theta_2 cos\theta_3 - 0.133\, sin\theta_2 \\ p_y &= -d_5\, cos\theta_2 sin(\theta_3+\theta_4) + cos\theta_2 cos\theta_3 + 0.133\, cos\theta_2 \\ p_z &= -d_5\, cos(\theta_3+\theta_4) - sin\theta_3 + d_1 \end{aligned} \tag{3}$$

The end effector posture equation is:

$$\begin{aligned} p_x &= d_5\, sin\theta_2 sin(\theta_3+\theta_4) - sin\theta_2 cos\theta_3 - 0.133\, sin\theta_2 \\ p_y &= -d_5\, cos\theta_2 sin(\theta_3+\theta_4) + cos\theta_2 cos\theta_3 + 0.133\, cos\theta_2 \\ p_z &= -d_5\, cos(\theta_3+\theta_4) - sin\theta_3 + d_1 \end{aligned} \tag{4}$$

2.3. Dynamic Model of Manipulator

According to the analysis of manipulator joints [21], the five degrees of freedom of the manipulator are respectively the lifting base, waist rotation axis, arm joint, and wrist rotation. The lifting base only works when the length of the manipulator is insufficient, while the waist rotation and wrist rotation axis are tentatively defined as not playing a key role in the visual recognition and picking process; therefore, the manipulator is simplified as a two-joint manipulator with only a big arm and a small arm. Meanwhile, a dynamic model is established, as shown in Figure 5.

$$M(q)\ddot{q} + C(q,\dot{q})\dot{q} + G(q) = \tau \tag{5}$$

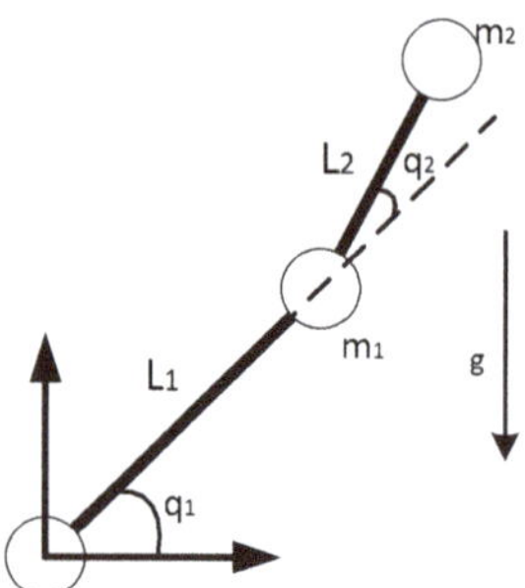

Figure 5. Simplified schematic diagram of two-joint manipulator.

Herein, $q = \left[q_1\ q_2\right]^T$, $\tau = [\tau_1\ \tau_2]^T$. In the formula, q, $\dot{q}$, $\ddot{q}$ are respectively the joint angle position, velocity, and acceleration vectors of the manipulator. $M(q) \in R^{n\times n}$ is a positive definite mass inertia matrix. $C(q,\dot{q}) \in R^{n\times n}$ represents the centrifugal force and Coriolis force. $G(q) \in R^n$ is gravity vector. $\tau \in R^n$ is the control moment input from each part of the joint. $\tau_d \in R^n$ is the unknown external disturbances. Therein, Formula (1) satisfies the following characteristics:

(1) Positive definite symmetry: For any q matrix, $M(q)$ is a positive definite symmetric.

(2) Boundedness: Matrix functions $M(q)$ and $C(q,\dot{q})$ are uniformly bounded for all q,$\dot{q}$.

(3) Skew symmetry: The matrix function $M(q) - 2C(q,\dot{q})$ is a skew symmetric matrix, i.e., for any vector α, there is $\alpha^T\left(M(q) - 2C(q,\dot{q})\right)\alpha = 0$.

3. Visual Servo Control

The visual servo control structure can be divided into position-based control and image-based control according to different feedback signals. The feedback signal of position-based control is defined in coordinate form in three-dimensional task space. Its basic principle is to estimate the target in a three-dimensional Cartesian coordinate system by extracting the image features and combining the geometric model of the known target, and then planning the trajectory and calculating the control quantity according to the position comparison between the end effector and the target of the manipulator. For the image-based control, the control quantity is calculated directly from the error signal in the image, and the task is completed by driving the manipulator [22]. In this paper, the image-based control method is used to output control signals after image calculation, and the sliding-mode control parameters are adaptively adjusted by the fuzzy neural network algorithm so as to improve the robustness and stability of the manipulator grasping.

3.1. *Visual Target Location*

According to the visual servo control system working process of fruit picking robots, the relationship between the position information of the robot and its joints should be determined first. The mapping relationship between the image plane coordinates and the three-dimensional coordinates in the real space can be established by the perspective model of the small hole so that the coordinate position of the target can be calculated [23]. Due to the vision camera adopted in this paper being the target localization of the monocular robot, its camera is installed at the end of the manipulator. According to the principle of the optical imaging of the camera, the model of monocular vision localization is established as shown in Figure 6, in which the mapping relation of two-dimensional image coordinates and coordinates in three-dimensional space is established by the transformed geometric coordinates.

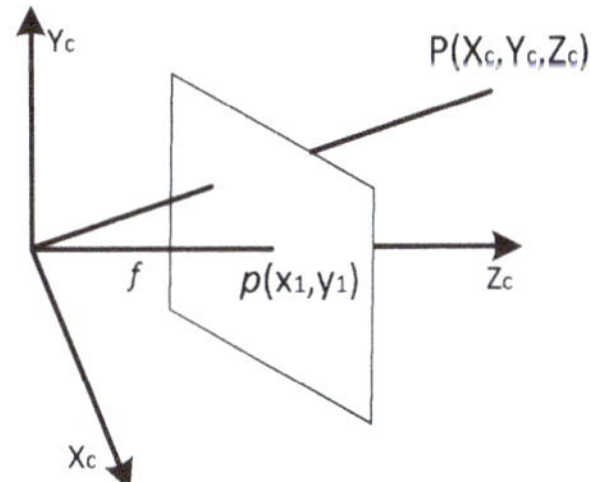

Figure 6. Visual model of camera imaging.

Here, we assume that the target fruit coordinate in the coordinate system of the manipulator is (X_a, Y_a, Z_a) and its coordinate in the camera coordinate system is (X_c, Y_c, Z_c). The relationship between the 3D coordinates (X_c, Y_c, Z_c) of the target in the camera coordinate system and its two-dimensional (2D) coordinates (X_1, Y_1) on the camera imaging plane is obtained.

$$\begin{cases} X_1 = \frac{X_c f}{Z_c} \\ Y_1 = \frac{Y_c f}{Z_c} \end{cases} \tag{6}$$

In the formula, f is the camera focal length. The coordinate transformation formula between the camera coordinate system and manipulator coordinate system can be obtained by the geometric analytic method shown in Figure 2.

$$\begin{cases} X = \sqrt{\frac{L_2^2+L_3^2-2L_2L_3\cos(\pi-\theta_3)-(L_2\sin\theta_2+L_3\sin(\theta_2+\theta_3))^2}{1+\tan^2\theta_1}} \\ Y = X\tan\theta_1 \\ Z = L_1+L_2\sin\theta_2+L_3\sin(\theta_2+\theta_3) \end{cases} \tag{7}$$

In the formula, L_1—Waist length

L_2—Big arm length
L_3—Small arm length

θ_1—one DOF joint angle of manipulator
θ_2—two DOF joint angle of manipulator
θ_3—three DOF joint angle of manipulator

From Formula (3), the relationship between the target fruit coordinate in the manipulator coordinate system (X_a, Y_a, Z_a), and the fruit coordinate in the camera coordinate system (X_c, Y_c, Z_c) can be obtained:

$$\begin{cases} X_c = X_a - X \\ Y_c = Y_a - Y \\ Z_c = Z_a - Z \end{cases} \tag{8}$$

The relationship between the robot angle change of each joint and the plane coordinate change of the target fruit imaging is obtained.

3.2. Visual Servo Control

Since the fruit picking robot works in unknown and uncertain environments and the visual servo control system has the characteristics of time-varying, strong coupling, and non-linearity, therefore, the visual servo control system based on images is used due to its advantages regarding insensitivity to errors. The structure of the image-based visual servo system is shown in Figure 7.

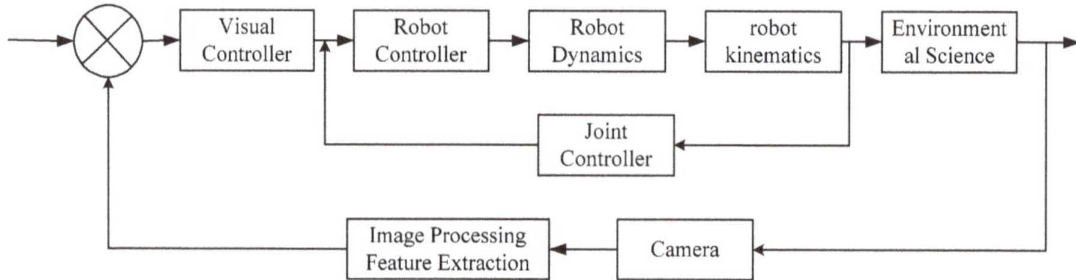

Figure 7. Image-based visual servo system structure.

In this system, the picking robot captures the target image with a camera installed by eye-in-hand mode, and the position information of the target fruit is obtained by an image processing system. The angle of each joint of the robot can be obtained by coordinate transformation. In order to improve the motion performance of the manipulator, a visual servo algorithm based on fuzzy neural network adaptive sliding mode control is adopted to overcome the uncertainty of the system so as to improve the robustness of the servo control of the manipulator.

4. Complex Fuzzy Neural Network Sliding Mode Control

4.1. Model Design of Sliding Mode Algorithms

Sliding-mode [24] variable structure control mainly uses a high-speed control rate under non-linear conditions. It can reach the predetermined sliding mode control track in a short time and move along the track according to the variable dynamic control system. The sliding mode control scheme is shown in Figure 8.

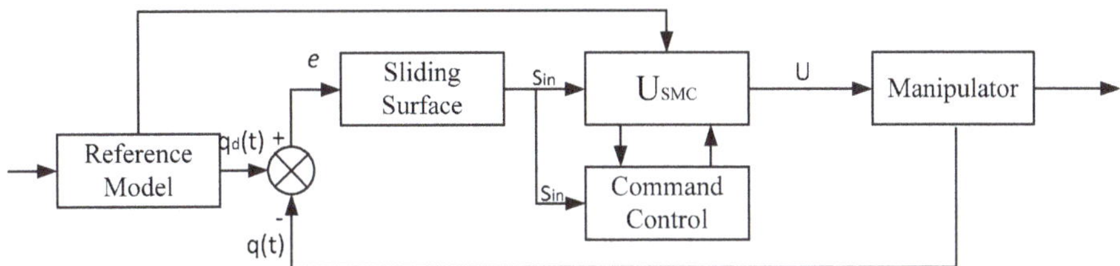

Figure 8. Structural diagram of the sliding-mode control (SMC) controller.

First, we set the ideal angle instruction to $q_d(t)$, and track the error of the sliding-mode controller:

$$e(t) = q(t) - q_d(t) \tag{9}$$

Then, we design the sliding surface as:

$$s(t) = \ddot{e} + \Lambda_1\dot{e} + \Lambda_2 e \tag{10}$$

Therein, $\boldsymbol{\Lambda} = \boldsymbol{diag}(\lambda_{i1}, \lambda_{i2}, \ldots, \lambda_{in}), \lambda_{ij} > 0, i = 1,2, j = 1,2,\ldots,n$.
Then, we set the Lyapunov function to:

$$V = \frac{1}{2}s^T M s \tag{11}$$

Therefore:

$$\dot{V} = \frac{1}{2}\dot{s}^T M s + \frac{1}{2}s^T \dot{M} s + \frac{1}{2}s^T M \dot{s} \tag{12}$$

Bring the characteristics of the manipulator dynamics equation into the formula:

$$\dot{V} = s^T M\dot{s} + s^T C\dot{s} = s^T\left(M\left(\dddot{q} - \dddot{q}_d + \Lambda_1\ddot{e} + \Lambda_2\dot{e}\right) + C\left(\ddot{e} + \Lambda_1\dot{e} + \Lambda_2 e\right)\right) \tag{13}$$

According to:

$$\tau. + \Lambda\tau = \Lambda U_{SMC} \tag{14}$$

Find:

$$\dot{V} = s^T(H + \Lambda U_{SMC}) \tag{15}$$

Therein:

$$\begin{aligned} H = M\left(\Lambda_1\ddot{e} + \Lambda_2\dot{e} - \dddot{q}_d\right) + C\left(\Lambda_1\dot{e} + \Lambda_2 e - \dddot{q}_d\right) - \left(\dot{M} + \Lambda M\right)\ddot{q} \\ -\left(\dot{C} + \Lambda C\right)\dot{q} - \left(\dot{G} + \Lambda G\right) \end{aligned} \tag{16}$$

Then, we design the SMC system as:

$$U_{SMC} = -\Lambda^{-1}(H + \eta sgn(s)) \tag{17}$$

Find:

$$\dot{V} = -\eta s^T sgn(s) = -\eta\|s\| \leq 0 \tag{18}$$

According to the LaSalle invariant set theorem, when $t \to \infty$, $s \to 0$, then $\tilde{q} \to 0$, $\dot{\tilde{q}} \to 0$.

4.2. Sliding-Mode Control Based on Complex Fuzzy Neural Network

The fuzzy neural network is a kind of special neural network that is a hybrid intelligent system formed by the combination of neural network and fuzzy logic. It combines the two kinds of techniques by combining the human-like reasoning of fuzzy systems with the learning and connection structure of neural networks [25,26]. In a nutshell, the fuzzy neural network (FNN) assigns a conventional neural network to fuzzy input signals and fuzzy weights. Its function is to use the fuzzy neural network structure to implement fuzzy logic reasoning.

In order to improve the control accuracy of the two-joint manipulator, a fuzzy neural network control algorithm, the fuzzy neural network (FNN) is constructed to optimize the conventional sliding mode control because of the chattering phenomenon in the conventional sliding-mode control [27–30]. The system consists of an input layer, an adaptive fuzzy rule layer, a rule layer, and an output layer. The structure of the system is shown in Figure 9. The input is a sliding surface, and the output is a control quantity. The following is the main flow chart of the FNN algorithm:

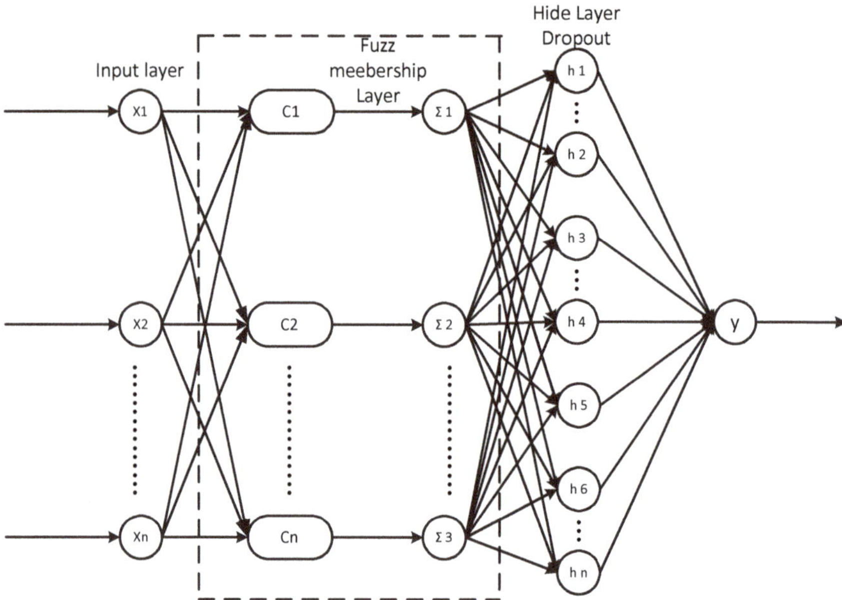

Figure 9. Structure of fuzzy neural network system.

Step 1: Input the layer data initialization of the fuzzy neural network, and transfer the sliding-mode vector variable $s_i(i = 1, 2, \ldots, n)$ to the next layer.

Step 2: The nodes in each layer represent the membership functions of the members, which use Gauss function as an input:

$$\mu_i^j(s_i) = exp\left[-\frac{\left(s_i - x_i^j\right)^2}{\left(\sigma_i^j\right)^2}\right], \ (i = 1, 2, \ldots, n; \ j = 1, 2, \ldots, m) \tag{19}$$

In the formula, n represents the number of input variables, and m is the number of member functions. x_i^j and σ_i^j represent the j level node of the i input from s_i to the Gauss function.

Step 3: The rule layer represents the fuzzy reasoning mechanism of the neural network, and the data results are obtained by multiplying the input signals of the layer. The formula is as follows:

$$g_k = \prod_{i=1}^{n} \omega_{ji}^k \mu_i^j(s_i) \tag{20}$$

Herein, g_k represents the Kth output of the rule layer.

Step 4: The fourth layer is the output layer, in which each node represents an output variable. It integrates the output into the sum of all the input signals. The execution output of this layer deblurring is as follows:

$$y_0 = \sum_{k=1}^{N_y} \omega_k^0 g_k \tag{21}$$

By adding FNN online cyclic training, inheriting the traditional sliding mode control system, and relaxing the requirement of system information, the SMC jitter problem can be eliminated. However, the traditional neural network training method mentioned above makes the learning speed of the algorithm slow and may fall into local minimum problem, etc. Therefore, a dropout layer is added on this basis to prevent over-fitting by randomly hiding layer nodes. The improved fuzzy neural network is expressed in vector form:

$$y = U_{NFNN}(e, x, \sigma, W, l, p) = Wg \tag{22}$$

In the formula, $\boldsymbol{W}$ is the weight matrix of the hidden layer and the output layer. $\boldsymbol{e}$ is the tracking error vector and the input vector of the system. $\boldsymbol{l}$ represents the fuzzy rules. $\boldsymbol{p}$ represents the dropout layer parameters of Bernoulli distribution.

Therefore, the equivalent control law of the optimal output estimation sliding surface is obtained.

$$U_{SMC} = U_{NN}^*(e, x^*, \sigma^*, W^*, l^*, p^*) + \varepsilon = W^* g^* + \varepsilon \tag{23}$$

Herein, ε represents the error vectors of the network. $\boldsymbol{W}^*, \boldsymbol{x}^*, \boldsymbol{\sigma}^*, \boldsymbol{l}^*, \boldsymbol{p}^*$ are the optimal parameters of network. Then, the sliding mode control law of the improved fuzzy neural network can be expressed as:

$$U(\mathrm{t}) = \hat{U}_{NN}\left(e, \hat{x}, \hat{\sigma}, \hat{W}, \hat{l}, \hat{p}\right) + U_r \tag{24}$$

Herein, $\hat{\boldsymbol{U}}_{NN}$ is the improved fuzzy neural network controller, which is used to estimate the sliding mode control law. $\boldsymbol{U}_r$ is an additional value of robustness, which ensures that the system would be on the sliding surface when $\mathrm{t} \geq 0$. $\hat{\boldsymbol{x}}, \hat{\boldsymbol{\sigma}}, \hat{\boldsymbol{W}}, \hat{\boldsymbol{l}}, \hat{\boldsymbol{p}}$ represent the neural network parameter estimation.

$$\widetilde{U} = U_{SMC} - U = W^* g^* + \varepsilon - \hat{W}\hat{g} - U_r = \widetilde{W} g^* + \hat{W}\widetilde{g} + \varepsilon - U_r \tag{25}$$

In the formula, $\widetilde{\boldsymbol{W}} = \boldsymbol{W}^* - \hat{\boldsymbol{W}}, \widetilde{\boldsymbol{g}} = \boldsymbol{g}^* - \hat{\boldsymbol{g}}, \widetilde{\boldsymbol{g}}$ can be expressed by Taylor expansion due to $\boldsymbol{g}$ being a non-linear activation function.

$$\widetilde{g} = g_x \widetilde{x} e + g_\sigma \widetilde{\sigma} e + o_m \tag{26}$$

Herein, $\widetilde{\boldsymbol{x}} = \boldsymbol{x}^* - \hat{\boldsymbol{x}}, \widetilde{\boldsymbol{\sigma}} = \boldsymbol{\sigma}^* - \hat{\boldsymbol{\sigma}}$ can be expressed as:

$$g^* = \hat{g} + g_x \widetilde{x} e + g_\sigma \widetilde{\sigma} e + o_m \tag{27}$$

Bringing Formula (22) into Formula (20), then we find:

$$\begin{aligned} \widetilde{U} &= W^* g^* + \varepsilon - \hat{W}\hat{g} - U_r \\ &= W^*(\hat{g} + g_x \widetilde{x} e + g_\sigma \widetilde{\sigma} e + o_m) + \varepsilon - \hat{W}\hat{g} - U_r \\ &= \left(W^* - \hat{W}\right)\hat{g} + \left(\widetilde{W} + \hat{W}\right) g_x \widetilde{x} e + \left(\widetilde{W} + \hat{W}\right) g_\sigma \widetilde{\sigma} e + \varepsilon - U_r + W^* o_m \\ &= \widetilde{W}\hat{g} + \hat{W} g_x \widetilde{x} e + \hat{W} g_\sigma \widetilde{\sigma} e - U_r + P \end{aligned} \tag{28}$$

Herein: $\boldsymbol{P} = \widetilde{\boldsymbol{W}} \boldsymbol{g}_x \widetilde{\boldsymbol{x}} \boldsymbol{e} + \widetilde{\boldsymbol{W}} \boldsymbol{g}_\sigma \widetilde{\boldsymbol{\sigma}} \boldsymbol{e} + \boldsymbol{W}^* \boldsymbol{o}_m + \boldsymbol{\varepsilon},\ \|\boldsymbol{P}\| < \boldsymbol{r},\ \boldsymbol{r}$ is a minimal positive number.

4.3. Stability Analysis

Next, we improve the fuzzy neural network sliding mode control algorithm according to the dynamics and kinematics equation of the manipulator to make the manipulator reach the prescribed sliding surface in the closed-loop control of the visual servo system, and reduce and eliminate the chattering problem of sliding-mode control, so as to ensure the long-term stability of the system.

Set:

$$\hat{W} = \eta_1\left(\hat{g}S^T\right)^T \tag{29}$$

$$\hat{x} = \eta_2\left(eS^T\hat{W}g_x\right)^T \tag{30}$$

$$\hat{\sigma} = \eta_3\left(eS^T\hat{W}g_\sigma\right)^T \tag{31}$$

$$U_r = \frac{\tau_0 sgn(S)}{\sum_{i=1}^{n} S_i^2} \tag{32}$$

Define the Lyapunov function as:

$$V = \frac{1}{2}S^T S + \frac{1}{2\eta_1}tr\left(\widetilde{W}\widetilde{W}^T\right) + \frac{1}{2\eta_2}tr\left(\widetilde{x}\widetilde{x}^T\right) + \frac{1}{2\eta_3}tr\left(\widetilde{\sigma}\widetilde{\sigma}^T\right) \tag{33}$$

Find the derivation:

$$= \frac{1}{2}S^T\dot{S} - \frac{1}{\eta_1}\mathrm{tr}\left(\widetilde{W}\widetilde{W}^T\right) - \frac{1}{\eta_2}\mathrm{tr}\left(\widetilde{x}\widetilde{x}^T\right) - \frac{1}{\eta_3}\mathrm{tr}\left(\widetilde{\sigma}\widetilde{\sigma}^T\right) \tag{34}$$

Since $\dot{S} = \widetilde{U}$, we introduce Formulas (6)–(15) and (6)–(17) and find:

$$\begin{aligned} \dot{V} &= S^T\left[\widetilde{W}\hat{g} + \hat{W}g_x\widetilde{x}e + \hat{W}g_\sigma\widetilde{\sigma}e - U_r + P\right] \\ &\quad -\tfrac{1}{\eta_1}\mathrm{tr}\left(\widetilde{W}\widetilde{W}^T\right) - \tfrac{1}{\eta_2}\mathrm{tr}\left(\widetilde{x}\widetilde{x}^T\right) - \tfrac{1}{\eta_3}\mathrm{tr}\left(\widetilde{\sigma}\widetilde{\sigma}^T\right) \\ &= \mathrm{tr}\left(\widetilde{W}\left(\hat{g}S^T - \tfrac{1}{\eta_1}\widetilde{W}^T\right)\right) + \mathrm{tr}\left(eS^T\hat{W}g_x - \tfrac{1}{\eta_2}\widetilde{x}^T\right) + \mathrm{tr}\left(eS^T\hat{W}g_\sigma - \tfrac{1}{\eta_3}\widetilde{\sigma}^T\right) + S^T(-U_r + P) \end{aligned} \tag{35}$$

According to Formulas (24), (27) and (30), we transform $\dot{\mathbf{V}}$ to:

$$\dot{V} = S^T(-U_r + P) \leq ||S^T|| \times || - U_r + P|| = -\alpha||S^T|| \leq 0 \tag{36}$$

Herein, α is a positive, sliding-mode control system based on the improved fuzzy neural network, and is asymptotically stable according to the Lyapunov stability principle. When $t \to \infty, s \to 0$.

5. Simulation Analysis and Experiment

The adaptive sliding-mode controller is simulated by matlab. The AC servo motor is used in the robot joint. The corresponding parameters are set according to the physical parameters of the picking robot arm as shown in Table 2.

$$H = \begin{bmatrix} \alpha + 2\varepsilon\cos(q_2) + 2\eta\sin(q_2) & \beta + \varepsilon\cos(q_2) + \eta\sin(q_2) \\ \beta + \varepsilon\cos(q_2) + \eta\sin(q_2) & \beta \end{bmatrix}$$

$$C = \begin{bmatrix} (-2\varepsilon\sin(q_2) + 2\eta\cos(q_2))\dot{q_2} & (-\varepsilon\sin(q_2) + \eta\cos(q_2))\dot{q_2} \\ (\varepsilon\sin(q_2) - \eta\cos(q_2))\dot{q_1} & 0 \end{bmatrix}$$

$$G = \begin{bmatrix} \varepsilon e_2\cos(q_1 + q_2) + \eta e_2\sin(q_1 + q_2) + (\alpha - \beta + e_1)e_1\cos q_1 \\ \varepsilon e_2\cos(q_1 + q_2) + \eta e_2\sin(q_1 + q_2) \end{bmatrix}$$

Table 2. Structure parameters of picking robot.

m_1	10.8 kg	l_{ce}	0.42 m
l_1	0.58 m	I_e	0.4 kg
l_{c1}	0.29 m	δ_e	0
I_1	0.9 kg	e_1	–0.58
m_e	8.2 kg	e_2	9.81

Herein, α, β, η, ε is constant. $\alpha = I_1 + m_1 l_{c1}^2 + I_e + m_e l_{ce}^2 + m_e l_1^2$, $\beta = I_1 + m_e l_{ce}^2$, $\varepsilon = m_e l_1 l_{ce} cos(\delta_e)$, $\eta = m_e l_1 l_{ce} sin(\delta_e)$. Take $\boldsymbol{a} = \begin{bmatrix} \alpha & \beta & \varepsilon & \eta \end{bmatrix}^{\mathrm{T}}$, $\hat{\boldsymbol{a}}$ is an estimation of $\boldsymbol{a}$, set $\widetilde{\boldsymbol{a}} = \hat{\boldsymbol{a}} - \boldsymbol{a}$, and $\boldsymbol{a}$ is constant, so $\dot{\widetilde{\boldsymbol{a}}} \approx \dot{\hat{\boldsymbol{a}}}$.

Set the angle instruction of joint 2 as $q_{d1} = sin(2\pi t)$, $q_{d2} = sin(2\pi t)$, $\dot{q}_{d1}$ and $\dot{q}_{d2}$ are respectively the angular velocities of joints 1 and 2. Take $\boldsymbol{\Lambda} = \begin{bmatrix} 15 & 0 \\ 0 & 15 \end{bmatrix}$, $\boldsymbol{\Lambda}_1 = \begin{bmatrix} 5 & 0 \\ 0 & 5 \end{bmatrix}$, $\boldsymbol{\Lambda}_2 = \begin{bmatrix} 50 & 0 \\ 0 & 50 \end{bmatrix}$, $\boldsymbol{K}_{\mathrm{d}} = \begin{bmatrix} 100 & 0 \\ 0 & 100 \end{bmatrix}$. Set the gain of the adaptive fuzzy neural network to $\eta_1 = 20$; $\eta_2 = 5$; $\eta_3 = 2$.

Firstly, the tracking curves of angular position and velocity of joints 1 and 2 are obtained by using sliding mode control without adding the fuzzy neural network algorithm, as shown in Figures 10–12.

From Figure 10, it can be seen that the angular position switching motion of common SMC joints without the fuzzy neural network is continuous and smooth. From Figure 12, it can be seen that joint 2 is more likely to be disturbed due to the need to deal with the conductive force from joint 1, so the output buffeting is relatively large.

The simulation results of a manipulator system with an improved fuzzy neural network sliding-mode control algorithm are shown in Figures 13–15. From Figure 15, it can be seen that compared with the traditional SMC scheme, the proposed fuzzy neural network sliding-mode control algorithm effectively eliminates the control chattering phenomenon. Meanwhile, the angular velocity tracking of joint 1 and joint 2 is more robust. It realizes high-precision control without detailed system information, and embodies the superiority of the improved fuzzy neural sliding-mode control algorithm.

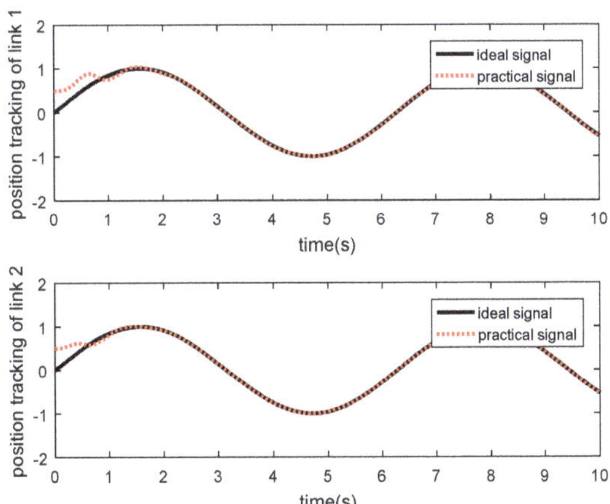

Figure 10. Angle tracking of joints 1 and 2.

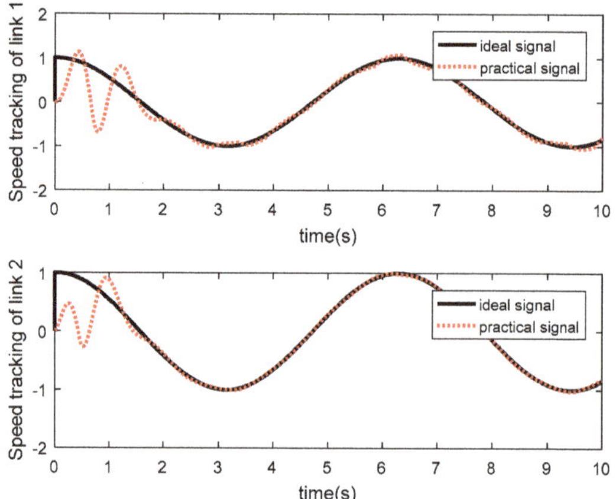

Figure 11. Angular velocity tracking of joints 1 and 2.

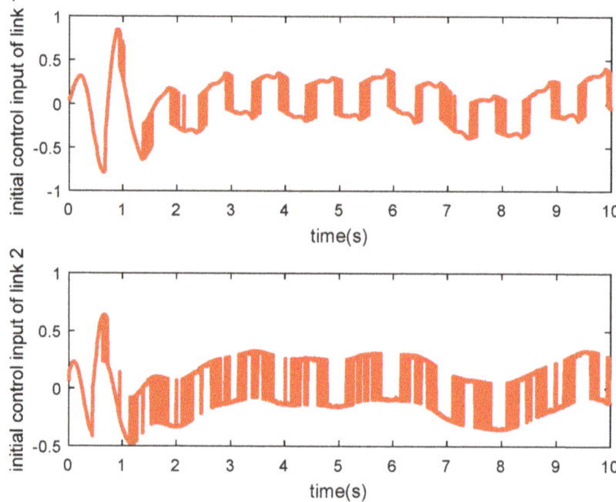

Figure 12. Joint 1 and 2 control inputs without neural network sliding mode.

In order to further verify the effectiveness of the sliding-mode algorithm of the fuzzy neural network system, the self-developed picking robot system is shown in Figure 16. The control system of the manipulator that is used in the experiment is an industrial computer with an Intel processor. The motion control system of the manipulator is connected with the host computer through serial communication, and the C control code generated by Matlab is imported into the control system.

The control instruction is generated according to the results of image recognition, so we process the control instructions by the sliding-mode algorithm of the fuzzy neural network and dynamic joint angle, output the dynamic joint angle, and bring into the kinematics equations related to the manipulator and output manipulator motor control instruction so as to control the manipulator gripper to reach the target position. Intensive experiments are performed on the apple-picking robot using the conventional PID algorithm, the sliding-mode control algorithm, and the improved fuzzy neural network sliding mode control on the robot arm servo control in the same experimental environment. Comparing the above three different algorithms on the picking-robot system on the grab speed and the grab accuracy, Table 3 is the data comparison using different algorithms.

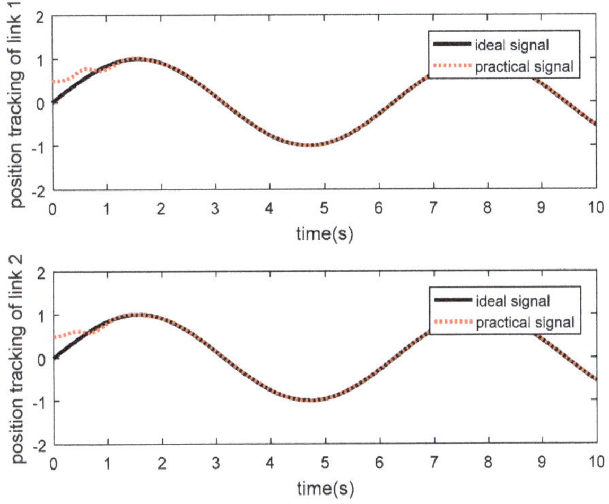

Figure 13. Position tracking of joints 1 and 2.

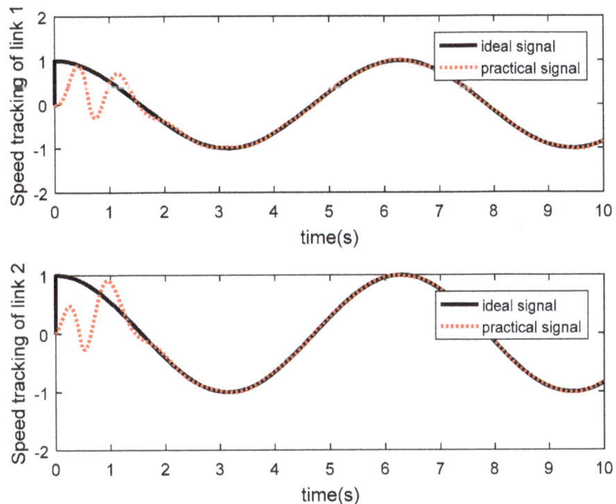

Figure 14. Angular velocity tracking of joints 1 and 2.

Table 3. Experimental results.

Algorithm	Success Rate	Starting Position to the End of the Crawl Time(s)
Conventional PID algorithm	88.2%	15.8
Sliding mode control algorithm	74.0%	13.0
improved fuzzy neural network sliding mode control algorithm	91.2%	13.8

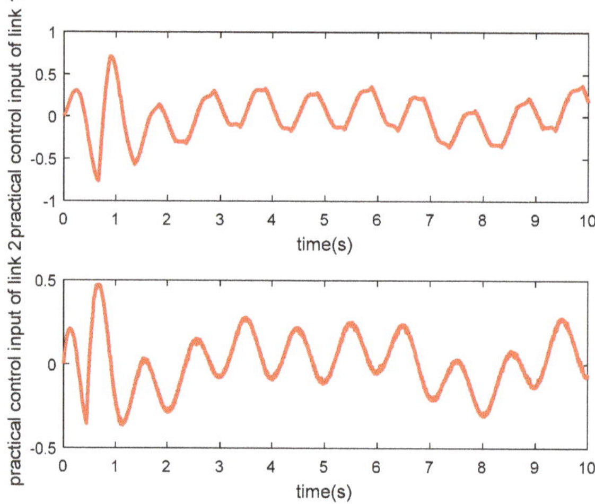

Figure 15. Control input of joints 1 and 2 after adopting the intelligent control scheme.

(**a**) (**b**)

Figure 16. (**a**) Representing the movement process of the robot arm of apple-picking robot; (**b**) Representing that the picking machine recognizes the target and controls the robot arm to grasp the target through the algorithm presented in this paper.

Through the comparison of experimental data, it can be known that the conventional PID algorithm captures a long time, while the sliding mode control algorithm can effectively reduce the grab time, but there is a lower catch success rate, and the improved fuzzy neural network algorithm is adopted. The sliding mode control algorithm not only improves the grabbing efficiency, but also greatly improves the success rate of the capture. It is shown in the result of the visual servo motion of the manipulator that the fruit-picking robot based on the improved fuzzy neural network sliding-mode algorithm has good stability and robustness. Figure 14 shows that the fruit-picking robot completes the visual servo control process of the manipulator using an improved algorithm.

6. Conclusions

In this paper, based on the self-developed apple-picking robot, the kinematics and dynamics equations of the robot are established. The image-based visual servo control method is adopted to control the manipulator in order to improve the grasping accuracy in the picking process. Then, the joint control performance of the control system has been improved by the adaptive fuzzy neural network sliding-mode control algorithm. Finally, the superiority of the algorithm is verified by simulation and experiment. The results show that the method improves the shortcomings of the common sliding mode control for the training control jitter in the actual robotic arm visual servo system to a certain extent, and improves the stability of the servo control.

Due to the low stability of the mechanical design of the current picking robot end effector, the existing mechanical structure of the picking robot will be improved in future work, and continue to verify the superiority of the current algorithm on the basis of the new terminal actuator, and complete the picking experiment in the real environment of the apple-picking robot.

Author Contributions: W.C. conceived and designed the experiments; W.C. performed the experiments; T.X. contributed reagents/materials/analyzed the data; T.X. and J.L. contributed reagents/materials/analysis tools; T.X. and M.W. wrote the paper; D.Z. analyzed the data.

Funding: This research is supported by the "Six talent peaks" high-level talents projection Jiangsu Province (2016-GDZB-021), Postgraduate Research & Practice Innovation Program of Jiangsu Province (KYCX18_2332), Project funded by China Postdoctoral Science Foundation (2016M601691), Industrial Foresight Project of Zhenjiang City (GY2018018).

Conflicts of Interest: The authors declare no conflict of interest.

References

1. Dong, G.; Zhu, Z.H. Position-based visual servo control of autonomous robotic manipulators. *Acta Astronaut.* **2015**, *115*, 291–302. [CrossRef]
2. Zhang, D.; Wei, B. A review on model reference adaptive control of robotic manipulators. *Ann. Rev. Control* **2017**, *43*, 188–198. [CrossRef]
3. Ganapathy, S.R.; Elumalai, V.K.; Karuppusamy, S. Uniform ultimate bounded robust model reference adaptive PID control scheme for visual servoing. *J. Frankl. Inst.* **2017**, *354*, 1741–1758. [CrossRef]
4. Ma, Z.; Sun, G. Dual terminal sliding mode control design for rigid robotic manipulator. *J. Frankl. Inst.* **2018**, *355*, 9127–9149. [CrossRef]
5. Li, X.; Cheah, C.C. Adaptive neural network control of robot based on a unified objective bound. *IEEE Trans. Control Syst. Technol.* **2013**, *22*, 1032–1043.
6. Krishna, S.; Vasu, S. Fuzzy PID based adaptive control on industrial robot. *J. Mater. Today* **2018**, *5*, 13055–13060. [CrossRef]
7. Namvar, M.; Aghili, F. Adaptive force-motion control of coordinated robots interacting with geometrically unknown environments. *IEEE Trans. Robot.* **2005**, *21*, 678–694. [CrossRef]
8. Zhao, Q.B.; Zhao, D.; Ji, W. Design of Visual Servo Control System for Picking Robot. *J. Agric. Mach.* **2009**, *40*, 152–156.
9. Cui, N.; Wang, B.; Mao, N. A Sliding Mode Control Method for Manipulator Terminal. *Aerosp. Control* **2018**, *173*, 34–40.
10. Haykin, S. *Neural Networks—A Comprehensive Foundation*; Prentice Hall: Upper Saddle River, NJ, USA, 1998; pp. 71–80.
11. Ali, A.A.; Lateef, R.A.R.; Saeed, M.W. Intelligent tuning of vibration mitigation process for single link manipulator using fuzzy logic. *Eng. Sci. Technol.* **2017**, *20*, 1233–1241. [CrossRef]
12. Yen, V.T.; Wang, Y.N.; Cuong, P.V. Robust adaptive sliding mode control for industrial robot manipulator using fuzzy wavelet neural networks. *Int. J. Control Autom. Syst.* **2017**, *15*, 2930–2941. [CrossRef]
13. Mehta, S.S.; Ton, C.; Rysz, M.; Kan, Z.; Doucette, E.A.; Curtis, J.W. New Approach to Visual Servo Control using Terminal Constraints. *J. Frankl. Inst.* **2019**, in press. [CrossRef]
14. Mehta, S.S.; MacKunis, W.; Burks, T.F. Nonlinear Robust Visual Servo Control for Robotic Citrus Harvesting. *IFAC Proc. Vol.* **2014**, *47*, 8110–8115. [CrossRef]
15. He, W.; Dong, Y. Adaptive Fuzzy Neural Network Control for a Constrained Robot Using Impedance Learning. *IEEE Trans. Neural Netw. Learn. Syst.* **2018**, *29*, 1174–1186. [CrossRef] [PubMed]
16. Er, M.J.; Gao, Y. Robust adaptive control of robot manipulators using generalized fuzzy neural networks. *IEEE Trans. Ind. Electron.* **2003**, *50*, 620–628.
17. Camci, E.; Kripalani, D.R.; Ma, L. An aerial robot for rice farm quality inspection with type-2 fuzzy neural networks tuned by particle swarm optimization-sliding mode control hybrid algorithm. *Swarm Evol. Comput.* **2018**, *41*, 1–8. [CrossRef]

18. Adhikary, N.; Mahanta, C. Sliding mode control of position commanded robot manipulators. *Control Eng. Pract.* **2018**, *81*, 183–198. [CrossRef]
19. Salgado, I.; Chairez, I. Adaptive unknown input estimation by sliding modes and differential neural network observer. *IEEE Trans. Neural Netw. Learn. Syst.* **2017**, *29*, 3499–3509.
20. Srivastava, N.; Hinton, G.; Krizhevsky, A.; Sutskever, I.; Salakhutdinov, R. Dropout: A simple way to prevent neural networks from overfitting. *J. Mach. Learn. Res.* **2014**, *15*, 1929–1958.
21. Wei, C.; Hao, L.; Yang, T.; Liu, J. Trajectory Optimization of Electrostatic Spray Painting Robots on Curved Surface. *Coatings* **2017**, *7*, 155.
22. Judith, M.; Frese, U.; Thomas, R. Grab a mug—Object detection and grasp motion planning with the Nao robot. In Proceedings of the IEEE-RAS International Conference on Humanoid Robots, Atlanta, GA, USA, 15–17 October 2013.
23. Yang, M.Y.; Ge, Y.Z. Robot visual servo control based on BP neural network. *Comput. Appl.* **2017**, *37* (Suppl. 2), 279–282, 297.
24. Liu, J.K. *MATLAB Simulation of Sliding Mode Variable Structure Control*, 3rd ed.; Tsinghua University Press: Beijing, China, 2012; pp. 179–190.
25. Hsieh, J.G.; Jeng, J.H.; Lin, Y.L.; Kuo, Y.S. Single index fuzzy neural networks using locally weighted polynomial regression. *Fuzzy Sets Syst.* **2019**, *368*, 82–100. [CrossRef]
26. De Campos Souza, P.V.; Torres, L.C.B.; Guimaraes, A.J. Data density-based clustering for regularized fuzzy neural networks based on nullneurons and robust activation function. In *Soft Computing*; Springer: Berlin/Heidelberg, Germany, 2019.
27. Chen, W.; Wang, X.; Liu, H.; Tang, Y.; Liu, J. Optimized Combination of Spray Painting Trajectory on 3D Entities. *Electronics* **2019**, *8*, 74. [CrossRef]
28. Mehta, S.S.; Mackunis, W.; Burks, T.F. Robust Visual Servo Control in The Presence of Fruit Motion for Robotic Citrus Harvesting. *Comput. Electron. Agric.* **2016**, *123*, 362–375. [CrossRef]
29. Wai, R.J.; Muthusamy, R. Fuzzy-neural-network Inherited Sliding-mode Control for Robot Manipulator including Actuator Dynamics. *IEEE Trans. Neural Netw. Learn. Syst.* **2013**, *24*, 274–287. [PubMed]
30. Xiao, Z.; Li, T.; Li, Z. A novel single fuzzy approximation based adaptive control for a class of uncertain strict-feedback discrete-time nonlinear systems. *Neurocomputing* **2015**, *167*, 179–186. [CrossRef]

Article

Optimized Combination of Spray Painting Trajectory on 3D Entities

Wei Chen [1,2,*], Xinxin Wang [1], Hao Liu [1], Yang Tang [3] and Junjie Liu [1]

[1] School of Electronics and Information, Jiangsu University of Science and Technology, Zhenjiang 212003, China; 172030030@stu.just.edu.cn (X.W.); directionod@aliyun.com (H.L.); junjiel_mtr@163.com (J.L.)
[2] School of Automation, Southeast University, Nanjing 210096, China
[3] School of Science, Jiangsu University, Zhenjiang 212013, China; ty800117@ujs.edu.cn
* Correspondence: cwchenwei@aliyun.com

Received: 27 November 2018; Accepted: 4 January 2019; Published: 9 January 2019

Abstract: In this research, a novel method of space spraying trajectory optimization is proposed for 3D entity spraying. According to the particularity of the three-dimensional entity, the finite range model is set up, and the 3D entity is patched by the surface modeling method based on FPAG (flat patch adjacency graph). After planning the spray path on each patch, the variance of the paint thickness of the discrete point and the ideal paint thickness is taken as the objective function and the trajectory on each patch is optimized. The improved GA (genetic algorithm), ACO (ant colony optimization), and PSO (particle swarm optimization) are used to solve the TTOI (tool trajectory optimal integration) problem. The practicability of the three algorithms is verified by simulation experiments. Finally, the trajectory optimization algorithm of the 3D entity spraying robot can improve the spraying efficiency.

Keywords: spray painting robot; FPAG; GA; ACO; PSO; TTOI problem

1. Introduction

With the development of the social economy and the improvement of life, people have higher requirements for product quality and product appearance. Moreover, some products with higher surface spraying quality, such as furniture, automobiles, and artworks, determine the quality of the product appearance and the competitiveness of products in the market. Therefore, people pay more and more attention to surface spraying technology [1]. The traditional spraying technology is that the spraying workers with spraying guns directly spray the workpiece to be sprayed. During spraying, the paint mist diffuses into the surrounding environment, which not only pollutes the environment, but also seriously harms the physical and mental health of spraying workers. To solve the problems caused by traditional spraying, automatic spraying systems have been developed. As a typical table of spraying automation equipment, a spraying robot has many advantages, such as good uniformity of the coating thickness, high repetitive positioning accuracy, wide applicability, and high efficiency. At the same time, spraying robots can free workers from a toxic, flammable, and explosive working environment [2].

In many fields of industrial production, the surface spraying of the workpiece is an essential step, and the types of workpiece modeling are becoming more and more abundant. During the spraying operation, the workpiece may have multiple spraying surfaces, and the curvature of each spraying surface is different. Usually, a three-dimensional (3D) entity modeling method is needed to form these parts, as shown in the 3D entity diagram in Figure 1. In recent years, due to the wide application of 3D solid modeling in various fields, scholars have made remarkable achievements in the research of 3D solid modeling [3–7]. The spraying of 3D solid modeling has a certain novelty because the spraying is more interested in the surface shape of the workpiece. For the research of this paper, we use the finite spraying modeling technology and the surface modeling method based on the plane patch adjacency graph (FPAG) for modeling. It should be pointed out that the 3D entities mentioned in this paper are not the same as polyhedra. A polygon is a spatial geometry surrounded by several planar polygons, each of which is a plane. However, each face of a 3D entity can be a free-form curved surface. In this sense, the polyhedron is only a subset of the 3D entity.

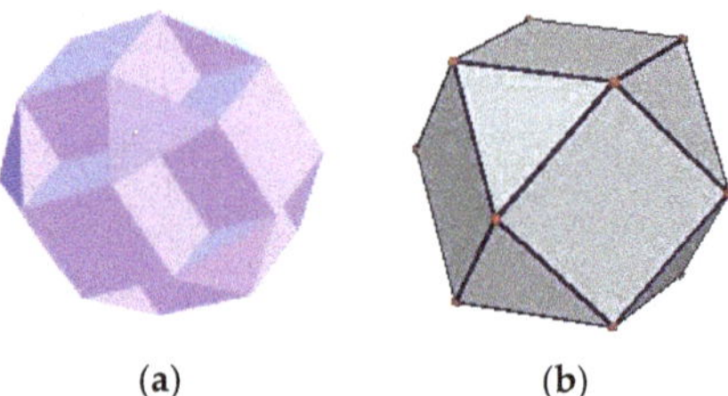

(a) **(b)**

Figure 1. (**a**) 3D entities with concave surfaces. (**b**) 3D entities with convex surfaces.

In addition, research on trajectory optimization for spray painting robots oriented to 3D entities is immature. Existing trajectory optimization methods can only be applied to 3D entities with convex surfaces, as shown in Figure 1b. For a 3D entity with concave surfaces (Figure 1a), the research on the trajectory optimization for the robot in this field is still a blank as the shape of the entity is complex and the robot is required to be extremely flexible in automatic spray painting operation. The trajectory optimization method for the spray painting robot introduced in this paper can only be applied to 3D entities with convex surfaces.

The idea of trajectory optimization for a spray painting robot oriented to a 3D entity is: First, a simple mathematical model of the paint deposition rate is established by the experimental method and the 3D entity is sliced by using the FPAG surface modeling method [8–10]. Secondly, after planning the painting path on each patch, the spray painting trajectory is optimized on each patch with the objective function of the variance of the paint thickness at discrete points and the ideal paint thickness. Finally, the spray painting trajectories on the individual patches are optimized, and the optimized trajectories of the spray painting robot on the 3D entities are formed eventually.

2. The Establishment of the Mathematical Model

The spraying mathematical model mainly includes the position and direction of the end-effector and the paint deposition rate model.

The establishment of the paint deposition rate model is an important problem in trajectory optimization for spray painting robots. Considering that the expressions of these models are rather complex, it is not applicable to build the mathematical model of the paint deposition rate on 3D entities. Therefore, in this paper, the specific mathematical expression of the finite range model takes the following factors into account: The distance and direction from the gun to the working surface, the curvature of the workpiece surface, and the angle of the paint gun (i.e., the cone angle corresponding to the paint flow).

Before building a finite-scope model, we made the following assumptions: The paint particles sprayed by the paint gun form a cone in space. Suppose the opening angle, φ, is half of the angle, φ, and the opening angle, $\varphi < 90°$. The definition of the opening angle, φ, can be seen in Figure 2.

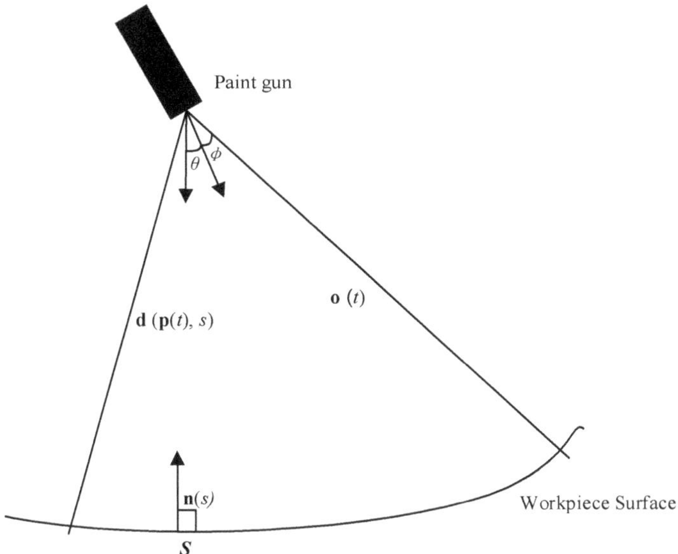

Figure 2. The finite range model.

Considering that the amount of paint to be obtained is smaller when the distance, $L(L > 0)$, from the point, $s(x, y, z) \in S$, on the surface to the gun increases. Additionally, the amount of paint obtained at the points on the surface is also decreased when the angle, $\theta(\theta < \varphi)$, is gradually increasing. The amount of the paint obtained at the point, s, on the workpiece surface can be expressed by the following equation:

$$\frac{c(\theta, \phi)}{L^2} \tag{1}$$

Among which:

$$c(\theta, \varphi)\begin{cases} > 0, & \theta < \varphi \\ = 0, & \theta \geq \varphi \end{cases} \tag{2}$$

$$L = \sqrt{(x - \mathbf{p}_x(t))^2 + (y - \mathbf{p}_y(t))^2 + (z - \mathbf{p}_z(t))^2} \tag{3}$$

$\mathbf{p}_x$, $\mathbf{p}_y$, and $\mathbf{p}_z$ express, respectively, the coordinates of the gun on the X-axis, Y-axis, and Z-axis. Assuming that the workpiece surface is a curved surface, the paint deposition rate of a point, $s(x, y, z) \in S$, on the surface is proportional to the inner product of the following two vectors (shown in Figure 2): First, the unit normal vector, $n(s)$, of the point. Second, the unit direction vector, $\boldsymbol{d}(\boldsymbol{p}(t), s)$, of the gun and the point, s. The function, $\boldsymbol{d}(\boldsymbol{p}(t), s)$, is defined as follows:

$$\boldsymbol{d}(\boldsymbol{p}(t), s) = \frac{(x - \boldsymbol{p}_x(t))\boldsymbol{i} + \left(y - \boldsymbol{p}_y(t)\right)\boldsymbol{j} + (z - \boldsymbol{p}_z(t))\boldsymbol{k}}{L} \tag{4}$$

where, $\boldsymbol{i}$, $\boldsymbol{j}$, and $\boldsymbol{k}$ denote unit vectors in the positive direction of the X, Y, and Z axis, respectively.

Based on the above assumptions, we can derive the paint deposition rate function of a point, s, on the surface of the workpiece:

$$\dot{f}(s, \boldsymbol{p}(t), t) = \left(\frac{c(\theta, \varphi)}{(x - \boldsymbol{p}_x(t))^2 + (y - \boldsymbol{p}_y(t))^2 + (z - \boldsymbol{p}_z(t))^2}\right) \cdot \boldsymbol{d}(\boldsymbol{p}(t), s) \cdot \boldsymbol{n}(s) \tag{5}$$

The variable, θ, in Equation (5) is not significant. In the equation, θ is the angle of the gun and a point, s, on the workpiece surface and the axis of the gun. It is related to the coordinates of the point, s, on the workpiece surface, the position, $\boldsymbol{p}(t)$, and direction, $\boldsymbol{o}(t)$, of the paint gun, which can be defined as:

$$\theta = \cos^{-1}(\boldsymbol{d}(\boldsymbol{p}(t), s) \cdot \boldsymbol{o}(t)) \tag{6}$$

The selection of function, $c(\theta, \varphi)$, depends on the basic spray characteristics of the gun and some parameter settings, such as the air pressure of the gun and the velocity of the paint flow. In the usual case, the function, $c(\theta, \varphi)$, reaches its maximum when $\theta = 0$ (that is, it just below the paint gun). When $\theta \to \varphi$, $c(\theta, \varphi) \to 0$. The following gives a concrete model of function, $c(\theta, \varphi)$:

$$c(\theta, \varphi) = \begin{cases} \alpha \frac{\cos(\theta) - \cos(\varphi)}{(1 - \cos(\varphi))^2} & \theta \leq \varphi \\ 0 & otherwise \end{cases} \tag{7}$$

The painting velocity of the paint gun and air pressure can be adjusted by changing the values of the parameters, α and φ, in the formula above, respectively. It can also be seen from the formula above that the values of the parameters, α and φ, are related to the maximum of the function, $c(\theta, \varphi)$.

The finite-range model here is circular on the plane corresponding to the paint distribution model. When the workpiece surface has a certain curvature, the paint distribution should become oval. For surfaces with small curvatures, the paint distribution, which is actually elliptical, can be approximated as a circular. In addition, the paint deposition rate model also illustrates that the total amount of paint sprayed to the workpiece surface (the total amount of paint sprayed from the gun) has nothing to do with the surface shape of the workpiece and the distance from the gun to the workpiece surface, which is in line with the actual situation.

3. Segmentation for 3D Entity

To obtain the optimal trajectory on the 3D solid surface, the first step is to model the workpiece surface. Second, due to the particularity of the spraying surface, trajectory optimization of the spraying robot is relatively difficult in a practical application. To obtain a good spraying effect, the finite range model modeling method is adopted. The third step is to use the FPAG surface modeling method to simplify the plane patch adjacency graph. As shown in Figure 3, the specific steps are as follows: (1) Divide the triangular grid of the 3D solid surface. (2) Set the maximum normal vector threshold and connect the triangle surface into a smaller flat area according to the triangle connection algorithm. (3) Each patch is approximately planar, and at least one side of each patch is part of the 3D solid ridge.

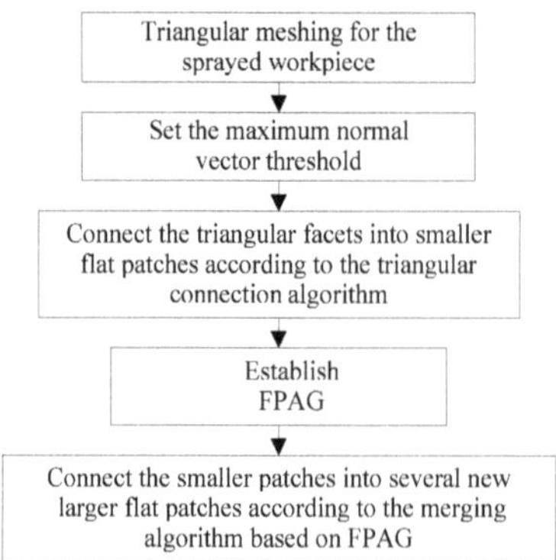

Figure 3. Step diagram of the surface modeling method based on the flat patch adjacency graph (FPAG).

4. Trajectory Optimization on Each Patch

The geometric properties of 3D entity surfaces are more complex compared with two-dimensional planar surfaces and regular curved surfaces. Therefore, to simplify the problem, we will describe a relatively simple and practical trajectory optimization method in the following. This method is fast and the process is simple. It can meet the actual needs completely.

The trajectory of a spray painting robot mainly consists of two factors: Path and velocity. In the spray painting process, the painting path can be obtained by determining the width of the overlapping areas formed by two paint strokes. Therefore, to determine the trajectory of a spray painting robot, we only need to determine the velocity of the paint gun and the width of the overlapping area formed by two paint strokes. x represents the distance from a point, s, to the first path in the painting radius, d represents the width of the overlapping area formed by two paint strokes, R represents the distance from the surface point to the spray direction, $o(t)$, O is the TCP (tool center point). Then, the paint thickness at point, s, is:

$$q_s(x) = \begin{cases} lq_1(x) & 0 \le x \le R-d \\ q_1(x) + q_2(x) & R-d < x \le R \\ q_2(x) & R < x \le 2R-d \end{cases} \tag{8}$$

$q_1(x)$ and $q_2(x)$ are the paint thickness at point s when spray painting on two adjacent paths, respectively. The formulas of $q_1(x)$ and $q_2(x)$ are:

$$q_1(x) = 2\int_0^{t_1} f(r_1)dt,\ 0 \le x \le R \tag{9}$$

$$q_2(x) = 2\int_0^{t_2} f(r_2)dt,\ R-d \le x \le 2R-d \tag{10}$$

among which, $t_1 = \sqrt{R^2 - x^2}/v$, $t_2 = \sqrt{R^2 - (2R-d-x)^2}/v$, $r_1 = \sqrt{(vt)^2 + x^2}$, $r_2 = \sqrt{(vt)^2 + (2R-d-x)^2}$. t_1 and t_2 represent half of the spray painting time that the gun paints in two adjacent painting paths at point s, respectively. r_1 and r_2 represent the distance from point s to the central projection point of TCP in two adjacent painting paths, respectively, t is the time that the gun moves from point O to point s'. s' is the projection of the point, s, on the painting path. The following expression can be obtained from (9) and (10):

$$q_s(x,d,v) = \frac{1}{v} J(x,d) \tag{11}$$

where, J is a function of x and d. To make the paint thickness on the surface as uniform as possible, the difference between the actual paint thickness and the variance of ideal paint thickness at point s are taken as the optimization objective function:

$$\min_{d\in[0,R],v} E_1(d,v) = \int_0^{2R-d} (q_d - q_s(x,d,v))^2 dx \tag{12}$$

In the equation above, q_d is the ideal paint thickness. Since the maximum paint thickness, $q_{\max}$, and the minimum paint thickness, $q_{\min}$, determine the uniformity of the paint thickness on the workpiece surface, and $q_{\max}$ and $q_{\min}$ also need to be optimized:

$$\min_{d\in[0,R],v} E_2(d,v) = (q_{\max} - q_d)^2 + (q_d - q_{\min})^2 \tag{13}$$

It can be obtained from Equations (11)–(13) that:

$$\min_{d\in[0,R],v} E(d,v) = \frac{1}{2R-d} E_1(d,v) + E_2(d,v) \tag{14}$$

Additionally, from Equations (9) and (10), the expressions of the maximum paint thickness and minimum paint thickness can be described as:

$$q_{\max} = \frac{1}{v} J_{\max}(d) \tag{15}$$

$$q_{\min} = \frac{1}{v} J_{\min}(d) \tag{16}$$

Let $\frac{\partial E(d,v)}{\partial v} = 0$, it can be obtained from Equations (11), (14)–(16) that:

$$v = \frac{\frac{1}{2R-d}\int_0^{2R-d} J^2(x,d)dx - J_{\max}^2(d) - J_{\min}^2(d)}{q_d\left[\frac{1}{2R-d}\int_0^{2R-d} J(x,d)dx + J_{\max}(d) + J_{\min}(d)\right]} \tag{17}$$

It can be seen that the spray painting rate, v, can be expressed as a function of the width, d, of the paint overlapping area formed by two painting strokes. Therefore, the minimum value of $E(d,v)$ is only related to d. The golden section method [11] can be used to obtain the optimization value, d, so that the optimized trajectory on each patch can be obtained too.

5. Tool Trajectory Optimal Integration on 3D Entity

After the track optimization of each patch, the optimal combination of trajectories connecting each patch should also be considered to speed up the spraying speed of the spraying robot. The first step is to transform and model the TTOI problem. TTOI problem is represented by the Hamiltonian diagram. In the second step, the corresponding optimization algorithm is used to solve the TTOI problem. The third step, through simulation and spray painting experiments, verification, and comparison, identifies the advantages and effectiveness of the algorithm.

5.1. The Transformation and Modeling of Tool Trajectory Optimal Integration

As is shown in the Figure 4, the TTOI (tool trajectory optimal integration) on each patch after the 3D entity segmentation is expressed [12]. To make the problem less complicated, the trajectories are considered as an edge. The ultimate purpose of the TTOI problem is to spray patches on the workpiece surface to make the spraying path of the robot the shortest. According to graph theory, a non-directional connection graph, G, is assumed (V, E, R, ω: $E \rightarrow Z^+$), among which V denotes the vertex set, E denotes the edge set, R denotes any subset of E, and ω denotes the weight of the edge (the length of the actual spray path). The problem of TTOI is to find a path passing all edges only once with the shortest distance in graph, G. Similar to the traveling salesman problem (TSP), which is a common problem in the optimization problem, the TTOI problem is also a typical NP (non-polynomial) problem.

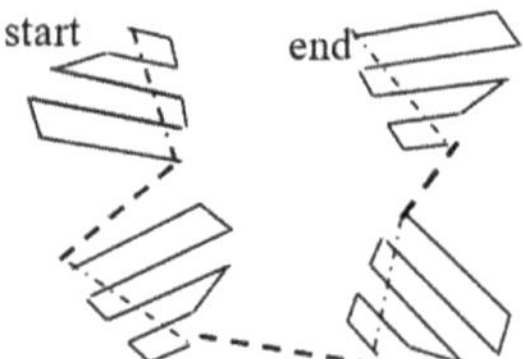

Figure 4. Tool trajectory optimal integration on each patch.

Suppose that D = $\{d_{ij}\}$ $(i, j = 1, 2, \ldots, n)$, the shortest distance between vertex i and vertex j, which are not on the same edge in graph G, the distance between the vertices can be calculated according to the Floyd algorithm. To make the problem less complicated, the TTOI problem can be expressed by the Hamiltonian method. As shown in Figure 5, a vertex is used to represent an edge of the original graph

G to form a complete Hamiltonian [12]: g (VH, EH, ωH), among which VH denotes the vertex set, EH denotes the edge set, and ωH denotes the weight of the edge and $\omega H \in$D. In the graph, g, the weight of each edge is not fixed. Its value is determined by the order of vertices on the same edge in the original graph, G. Suppose that the order of the vertices set, $VH = \{v_1, v_2 \ldots \ldots v_n\}$, in graph g is $T = (t_1, t_2 \ldots \ldots t_n)$ $t_i \in VH$ ($i = 1, 2, \ldots, n$), and the TTOI problem can be defined as Equation (18):

$$\min \overline{L} = \sum_{i=1}^{n} \omega_i + \sum_{j=1}^{n-1} \omega_j^{\mathrm{H}} \tag{18}$$

where ω_i is the weight of the edges in the primitive graph, G, corresponding to vertices, $t_1, t_2 \ldots \ldots t_n$, in graph g and ω_j^H denotes the weight of the edge in graph g. Since the weight, ω_i, of each edge in the original graph, G, is considered to be a fixed value in this problem, the above optimization problem can be reduced to Equation (12):

$$\min L = \sum_{j=1}^{n-1} \omega_j^{\mathrm{H}} \tag{19}$$

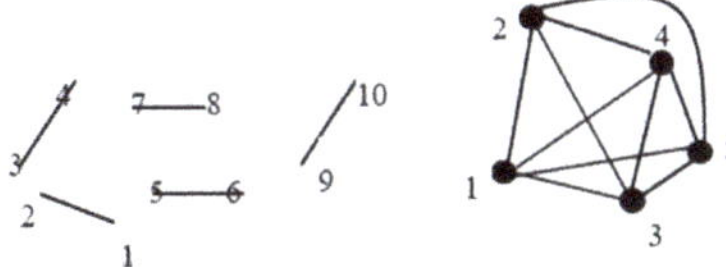

Figure 5. Transformation of the original graph, G, into the Hamiltonian graph, g.

The spraying robot is the most complex one in the control of the industrial robot because of its many parameters. Especially, the trajectory optimization of the spraying robot on the complex surface makes the actual operation difficult. Therefore, finding the arrangement of all vertices in the Hamilton diagram makes the path, L, of the painting robot the shortest, which becomes a TTOI problem. Because of the large number of parameters, spray robots are the most complex control of industrial robots. Especially, the trajectory optimization of the spraying robot on the complex surface makes the actual operation difficult. Therefore, finding the arrangement of all vertices in the Hamilton diagram makes the shortest path, L, of the painting robot a TTOI problem. To solve the TTOI problem, the improved genetic algorithm, ant colony algorithm, and particle swarm optimization (PSO) proposed in this paper can be used to optimize the trajectory of a spraying robot on complex surfaces. In the process of fragmentation, these intelligent algorithms can solve the trajectory optimization problem between patches. For the first time, these intelligent algorithms have been used for spraying complex surfaces. The previous links between patches are random combinations. Finally, the advantages and disadvantages of each algorithm are illustrated by experiments.

5.2. Solving the TTOI Problem with the Genetic Algorithm

Genetic algorithm (GA) is a method to search for the optimal solution by simulating the natural evolution process. Therefore, this algorithm has good effects on the NP (non-polynomial) problem in combinatorial optimization and can be used to solve the TTOI problem. According to the particularity of TTOI [13–15], GA needs a special individual code and crossover, mutation, and other genetic manipulation methods.

(1) Individual code: The length of the individual code is $|VH|$. Since each vertex in the Hamilton graph represents one edge of the original graph, G, to distinguish the start and end points of each edge. The individual code contains the binary code, P_{si}, representing the direction of each edge in the original graph, G.

(2) Fitness function: The values of fitness function are used to determine which individuals can enter the next round of evolution and which individuals need to be removed from the population.

To facilitate the selection operation in the genetic algorithm, the optimization of the minimum value is usually converted to optimization of the maximum value, and the fitness function can be taken as: $F = U - L$, where U should be selected as an appropriate number, to make the fitness of all individuals positive. In the process of population evolution, to select the individuals with high fitness, the population size is maintained as the value, P_{size}. According to the fitness function rule, the P_{size} individuals with the highest fitness are passed to the next generation.

(3) Crossover: Crossover is the process of exchanging the partial codes between two individuals with a certain probability to generate new individuals. Here, order crossover (OX) is used on P_i while two-point crossover is used on P_{si}. OX ensures that the original order of each vertex is almost the same when the effective sequence of the individual itinerary is modified [16]. The main idea of OX is: A conventional two-point crossover is performed, followed by an effective sequence modification of the individual itinerary. When modifying, the original relative access order of each point should be maintained as much as possible. Basic steps of OX are as follows:

(a) In the individual code strings, P_x and P_y, representing the spray painting order, the positions after the two loci, i and j, are randomly selected as the intersection. That is, each locus between the $(i + 1)$-th locus and the j-th locus is defined as an intersection area, and the contents of the intersection region are respectively memorized to W_x and W_y.

(b) According to the mapping relation in the intersection area, find all $P_{xq} - P_{xq}$ $(p = i + 1, i + 2, \ldots, j)$ loci q in the individual P_x and set them as vacancies. Find all $P_{xq} - P_{xq}$ $(p = i + 1, i + 2, \ldots, j)$ loci, r, in the individual, P_y, and set them as vacancies.

(c) The individuals, P_x, P_y, are left-shifted circularly until the first vacancy in the code string is moved to the left end of the intersection area. Then, all the vacancies are concentrated in the intersection area, and the original gene values in the intersection area are sequentially moved backward.

(d) Exchange the content in W_x and W_y and put them into the intersection area of individual P_x, P_y. The result is a new spray painting order.

(4) Mutation operation: P_i is subjected to inversion mutation to generate a new individual. A basic variation is applied to P_{si}, where one or more loci are randomly selected for individual code and the gene values of these loci are inverted.

Thus, the genetic algorithm of the TTOI problem is shown in Figure 6.

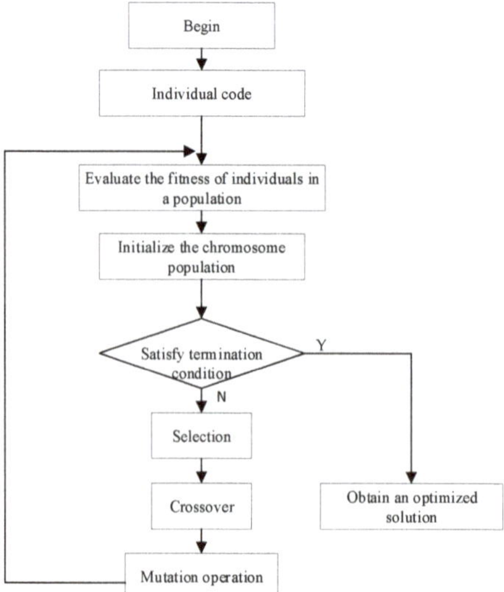

Figure 6. Flow chart of the GA (Genetic algorithm) of the TTOI (tool trajectory optimal integration) problem.

Taking 3D entities as spray objects, simulation experiments were carried out using genetic algorithm programming to verify the effectiveness of the TTOI problem. According to the segmentation method of the 3D entity, a 3D entity is divided into seven patches; that is, the individual code, P_i and P_{si}, are seven bits in the genetic algorithm. The parameters of the algorithm are as follows: Population size, P_{size} = 100; crossover probability, x_{rate} = 0.20; mutation probability, m_{rate} = 0.05; and the maximum number of evolutionary generation, T = 100. The corresponding evolutionary processes of different solutions are shown in Figure 7. As can be seen from the figure, the value of the objective function of the optimal individual decreases monotonously with the evolution process and eventually tends to be constant. After about 70 generations of evolution, the average fitness remains stable and the algorithm converges.

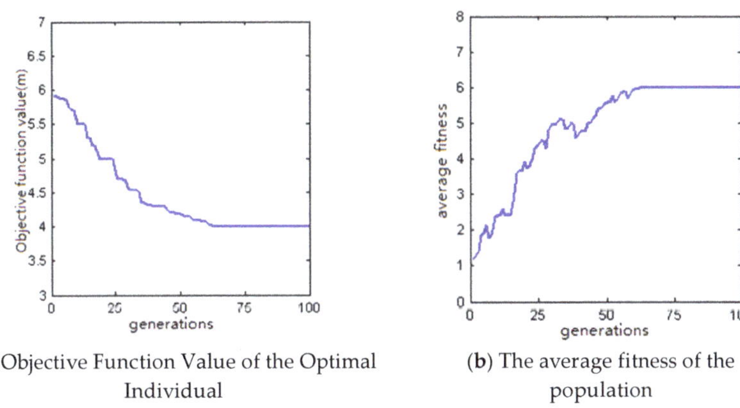

(**a**) Objective Function Value of the Optimal Individual (**b**) The average fitness of the population

Figure 7. The simulation result of solving the TTOI problem with GA.

5.3. Solving the TTOI Problem with the Ant Colony Algorithm

Ant colony optimization (ACO) is a probabilistic intelligent algorithm for finding the optimal path. It originated from the behavior of ants to find the path in the process of searching for food. It has strong anti-interference ability, strong compatibility, and other characteristics. The algorithm initializes the following individual information: Not visited vertices (NVV), not visited edges (NVE), visited edges (VE), and tour length (TL). Through the memory function of the population, individual information is constantly updated and adjusted. Taking the connection graph, G, shown in Figure 5 as an example, if the algorithm starts when the ant is at the vertex, 1, the initialization information is:

NVV [1] = {1,2,3,4,5,6,7,8,9,10}.
VV [1] = {}.
NVE [1] = {(1,2),(3,4),(5,6),(7,8),(9,10)}
VE [1] = {}.
TL [1] = 0.0

After time, Δt, the pheromone on trajectory (i, j) i adjusted as follows:

$$\tau_{ij}(t + \Delta t) = \rho\tau_{ij}(t) + \Delta\tau_{ij} \tag{20}$$

where ρ represents the volatilization rate of pheromone, $\tau_{ij}(t)$ represents the accumulation amount of pheromone on the track (i, j) at time t, $\Delta\tau_{ij}$ represents the increment of the pheromone on the trajectory (i, j) after the time, Δt, which can be calculated as follows:

$$\Delta\tau_{ij} = \sum_{k=1}^{m} \Delta\tau_{ij}^{k} \tag{21}$$

$\Delta\tau_{ij}^{k}$ denotes the pheromone on the trajectory (*i*, *j*) during the searching process of the *k*-th ant, the expression of which is:

$$\Delta\tau_{ij}^{k} = \begin{cases} \frac{Q}{TL[k]}, & ant\ k\ pass\ path\ (i,\ j) \\ 0, & else \end{cases} \tag{22}$$

Among which, Q is a constant. The pheromones on each trajectory during initialization are: $\Delta\tau_{ij} = 0$. At the time, t, the transition probability of an ant, k, from vertex x to other feasible vertices is:

$$p_{ij}^{k} = \begin{cases} \frac{(\tau_{ij}(t))^{\alpha}(\eta_{ij}(t))^{\beta}}{\sum\limits_{s \in alowed_k} (\tau_{is}(t))^{\alpha}(\eta_{is}(t))^{\beta}}, & j \in alowed_k \\ 0, & otherwise \end{cases} \tag{23}$$

where η_{ij} denotes the visibility on track (*i*, *j*), which reflects the degree of heuristic from vertex *i* to vertex *j*. Here, let $\eta_{ij} = 1/d_{ij}$, d_{ij} is the distance from vertex *i* to vertex *j*. The parameters, α and β, denote the influence weights of $\tau_{ij}(t)$ and $\eta_{ij}(t)$ on the whole transition probability. $alowed_k$ denotes the feasible neighborhood of the ant k at vertex i (end of the edge in the list NVE). Thus, the ant colony algorithm of the TTOI problem is shown in Figure 8.

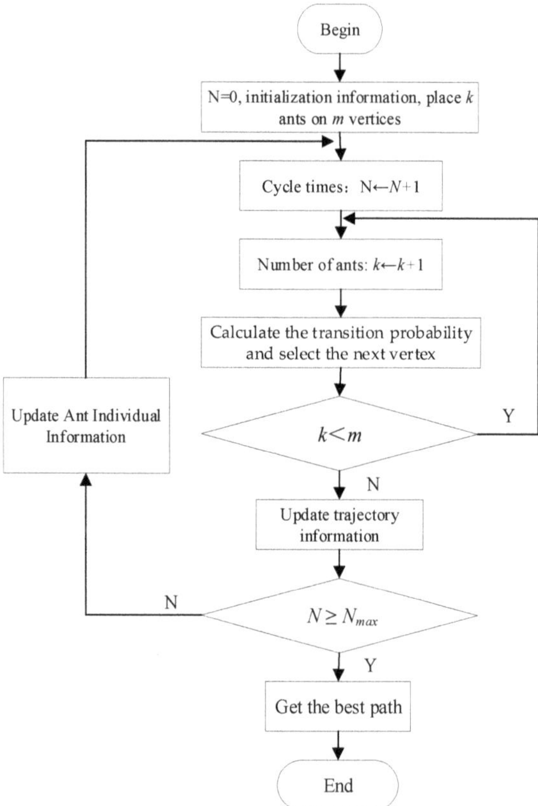

Figure 8. Flow chart of the ACO of the TTOI problem.

In the following, the validity of using ACO to solve the TTOI problem is verified by simulation experiments. Assuming that a 3D entity workpiece is divided into five patches, the number of edges in the connected graph, G, is five and the number of vertices is m = 10. The parameters of the algorithm are chosen as follows: $\alpha = 1$, $\beta = 5$, $\rho = 0.5$, $Q = 100$ and the maximum number of cycles is $N_{max} = 100$. Figure 9 shows the evolution of the optimal solution obtained from the algorithm. It can be seen from the figure that the length of the spray painting trajectory decreases monotonically with the evolutionary process. After about 70 generations of evolution, the length of the trajectory no longer changes and the algorithm converges.

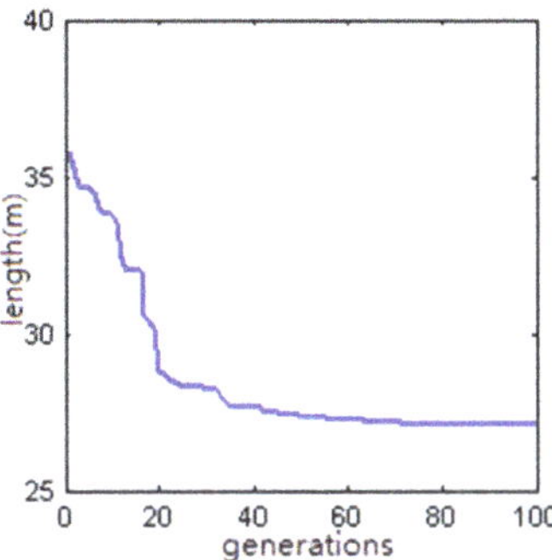

Figure 9. Simulation result of the ACO algorithm.

5.4. Solving the TTOI Problem with Particle Swarm Optimization

Particle swarm optimization (PSO) is easy to implement and there is no need to adjust a lot of parameters compared with other optimization algorithms. Also, it does not need gradient information, which is an effective tool to solve the optimal combination [17–19]. In the algorithm, each individual is a particle, and each particle represents a potential solution. Assuming that $z_i = (z_{i1}, z_{i2}, \ldots, z_{iD})$ is the D-dimensional position vector of the i-th particle, the current fitness value of z_i can be calculated according to the fitness function so that the position of the particle can be measured. The process of calculating the minimum length of the spray path length can be selected as the fitness function in the TTOI problem. $v_i = (v_{i1}, v_{i2}, \ldots, v_{iD})$ is the flying speed of particle i; that is, the distance of particle movement. $p_i = (p_{i1}, p_{i2}, \ldots, p_{iD})$ is the optimal position of the particle to date. $p_g = (p_{g1}, p_{g2}, \ldots, p_{gD})$ is the optimal position of the particle swarm to date. In each iteration, the velocity and position of the particles can be updated according to:

$$v_{id}^{k+1} = v_{id}^{k} + c_1 r_1 (p_{id} - z_{id}^{k}) + c_2 r_2 (p_{gd} - z_{id}^{k}) \tag{24}$$

$$z_{id}^{k+1} = z_{id}^{k} + v_{id}^{k+1} \tag{25}$$

where $i = 1, 2, \ldots, m$, $d = 1, 2 \ldots D$, r_1 and r_2 are random numbers between [0, 1], and c_1 and c_2 are learning factors. Thus, the particle swarm algorithm of the TTOI problem is shown in the Figure 10.

Assuming that the 3D entity artifact is divided into five patches, the number of edges in the connection graph, G, shown in Figure 5 is five and the number of vertices is m = 10. To guarantee the precision of the algorithm, the maximum number of cycles is $N_{max} = 100$. To ensure that the particle does not skip the optimal solution and can search the search space sufficiently, let $\varepsilon = 1000$. To ensure accuracy and reduce the amount of calculation, take the number of particles as 20. The learning factors, c_1 and c_2, can make the particles have self-summary and the ability to learn from the outstanding individuals in the group, to be close to the best in their own history and the history within the group. These two parameters have little effect on the convergence of the algorithm, but if we adjust these two parameters properly, we can reach the convergence faster. After adjusting the values of c_1 and

c_2 several times and analyzing the effect of c_1 and c_2 on the optimal fitness, we can conclude that $c_1 = c_2 = 2$ is a better choice for the TTOI problem. Figure 11 shows the evolution of the optimal solution obtained from the algorithm. It can be seen from the figure that the length of the spray painting trajectory decreases monotonously with the evolutionary process, and finally tends to a definite value. As the spray painting robot does not consider the obstacle avoidance during the working process, the environment information is known and is relatively simple, so the particle swarm algorithm converges faster. It can be seen from Figure 11 that after about 80 generations of evolution, the length of the spray painting trajectory does not change and the algorithm converges.

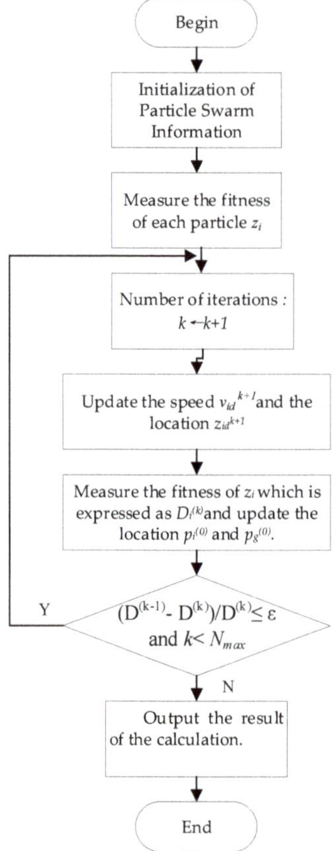

Figure 10. Flow chart of the PSO of the TTOI problem.

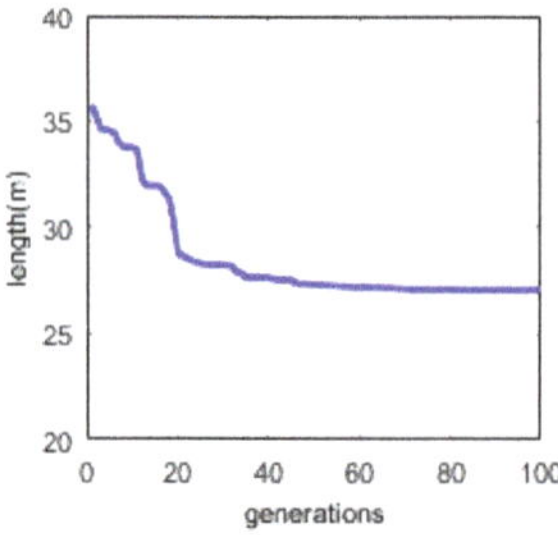

Figure 11. Simulation result of the PSO algorithm.

5.5. Comparison of Algorithms

In this paper, according to the built coating accumulation model: The finite range model, then FPAG (flat patch adjacency graph) is used to treat spraying workpiece, and the workpiece surface is divided into numerous patches. In each patch, the trajectory optimization algorithm in part 3 is used to obtain the trajectory. Finally, the genetic algorithm, ant colony algorithm, and particle swarm optimization algorithm are respectively used to obtain the optimal path to connect all patches to complete the spraying work. Assuming the ideal paint thickness of q_d = 50 μm, the maximum allowable deviation of paint thickness of q_w = 10 μm, the bottom radius of the cone paint sprayed by the gun is R = 60 mm. The cumulative rate of the paint is obtained from the spray test data on the plate [12]:

$$f(r) = \frac{1}{15}(R^2 - r^2)\mu\text{m/s} \tag{26}$$

After generating and optimizing the spray painting trajectories on the plate, the spray painting rate (at uniform speed) and the width of the overlapping area of the paint for each of the two spray strokes are obtained, which are v = 256.3 mm/s, d = 50.2 mm, respectively.

As shown in Figure 12, taking the experimental workpiece as an example, the workpiece can be regarded as a 3D entity. In the experiment, suppose the vertical distance from the gun to the workpiece surface is H, the bias algorithm can be used to obtain the spray path. Each spray parameter setting in the optimal algorithm of the spray painting trajectory are as follows: The ideal paint thickness is q_d = 50 μm, the maximum allowable deviation of thickness is q_w = ± 10 μm, the spray radius is R = 60 mm, the spray distance is H = 100 mm, and the velocity of the spray is v = 256.3 mm/s. The workpiece surface is divided into five patches after the modeling work and the path at the junction of the two patches is the PA-PA mode. GA, ACO, and PSO are used to carry out experiments when solving the TTOI. The parameters of the GA algorithm are set as: Population size is P_{size} = 100, crossover probability is x_{rate} = 0.20, mutation probability is m_{rate} = 0.05, and the maximum number of evolutionary generation is T = 100. The parameters of the ACO algorithm are set as: The number of ants is m = 10, parameter α = 1, parameter β = 5, the volatilization rate of the pheromone is ρ = 0.5, constant Q = 100, and the maximum number of iterations is N_{max} = 100. The parameters of the PSO algorithm are set as: The maximum speed threshold is ε = 1000, the number of particles is 20, learning factors are c_1 = c_2 = 2, and the maximum number of iterations is N_{max} = 100.

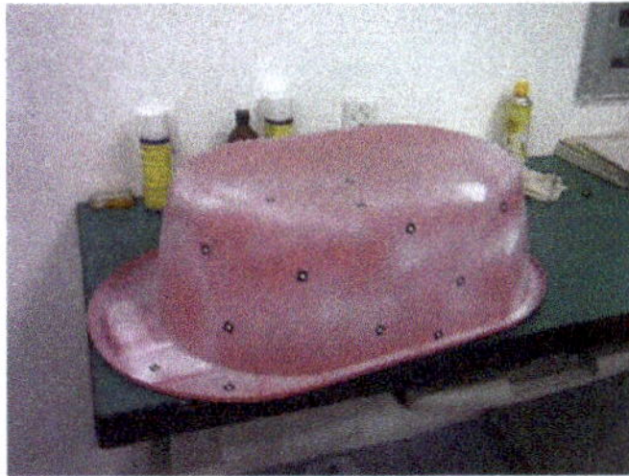

Figure 12. The experimental workpiece.

The self-developed offline programming system of the spray painting robot is used for the spray test [20,21]. The part of the optimized trajectories of different patches on the workpiece surface are shown in Figure 13. The paint thicknesses of 400 discrete points are measured by the paint thickness gauge after the spray painting operation. The paint thickness at the sample point after spray painting in the optimized trajectory is shown in Figure 14, among which the maximum paint thickness is q_{max} = 55.6 μm, and the minimum paint thickness is q_{min} = 45.5 μm. It can be seen that all the paint thicknesses at the sampling point are within the maximum allowable deviation of the paint thickness q_w, which is in line with the requirements of the spray quality.

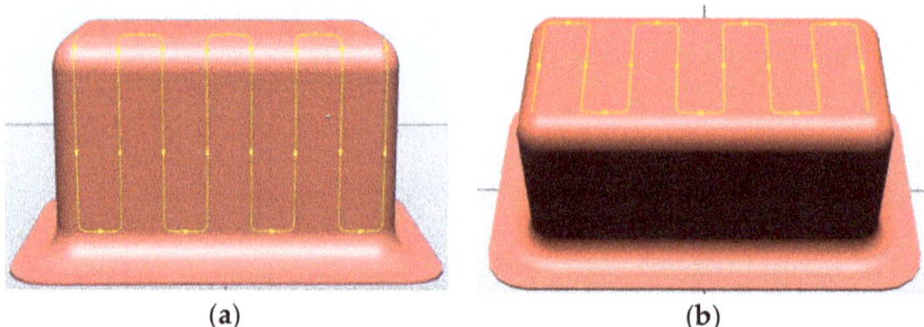

Figure 13. (a) Optimized trajectory at the side of the patch; (b) optimized trajectory at the top of the patch.

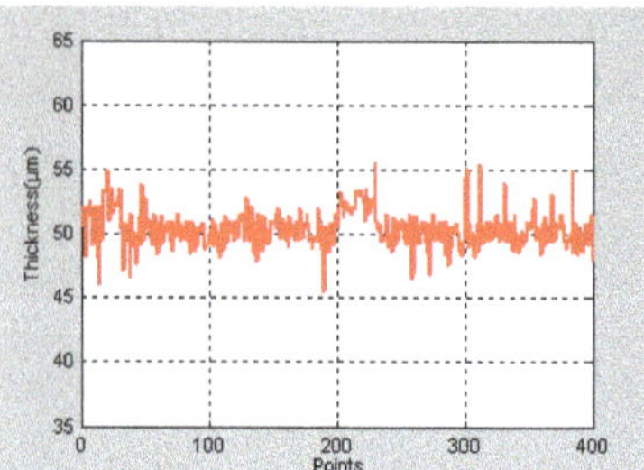

Figure 14. The paint thickness at the sampling points.

From the point of view of spray painting efficiency, the compared results of GA, ACO, PSO, and random combination are shown in Table 1. It can be seen from the results that the total length of the spray trajectory using the PSO algorithm is the shortest, the spray painting time is the least, and the execution time of the system operation is the longest, which is within the allowable range in the practical application. For the workpiece, the spray painting time was reduced by 23% compared to a random combination. The total trajectories of GA and ACO are shorter than those of the random combination, and the spray painting time are reduced by 16% and 20%, respectively. It should be noted that the workpiece used here is only divided into five patches. For the workpiece, which is more complex and has more patches, the advantages of using the PSO algorithm in saving painting time will be more obvious, but the execution time of the offline programming system will be longer. Therefore, if the real-time performance of the system operations can meet the practical application requirements, the PSO algorithm will be the best choice. Otherwise, the GA algorithm or ACO algorithm can be taken into consideration.

Table 1. The compared results of different algorithms.

	GA	ACO	PSO	Random Combination
Total length of Spray Path (m)	28.4	27.6	26.8	29.8
Spray painting Time of Robot (s)	94	89	86	112
Execution Time of Operation (s)	0.23	0.35	0.52	0.10

5.6. Spray Painting Experiment

As shown in Figures 15–18, taking the automotive body of a brand is as the paint objective. The thickness of the coating on each sampling point is shown in Figure 19. Each spray parameter settings in the optimal algorithm of the spray painting trajectory are as follows: The ideal paint thickness is q_d = 50 μm, the maximum allowable deviation of the thickness is $q_w = \pm 10$ μm, the spray radius is R = 60 mm, the spray distance is H = 80 mm, and the velocity of the spray is v = 389 mm/s. The left side of the car surface is divided into 15 patches and the trajectory of each patch is in a different color. GA, ACO, and PSO are used to carry out experiments when solving the TTOI. The parameters of the GA algorithm are set as: Population size is $P_{size} = 100$, crossover probability is $x_{rate} = 0.20$, mutation probability is $m_{rate} = 0.05$, and the maximum number of evolutionary generation is $T = 200$. The parameters of the ACO algorithm are set as: The number of ants is $m = 10$, parameter $\alpha = 1$, parameter $\beta = 5$, the volatilization rate of pheromone is $\rho = 0.5$, constant $Q = 100$, and the maximum number of iterations is $N_{max} = 200$. The parameters of the PSO algorithm are set as: The maximum speed threshold is $\varepsilon = 1000$, the number of particles is 20, learning factors $c_1 = c_2 = 2$, and the maximum number of iterations is $N_{max} = 200$.

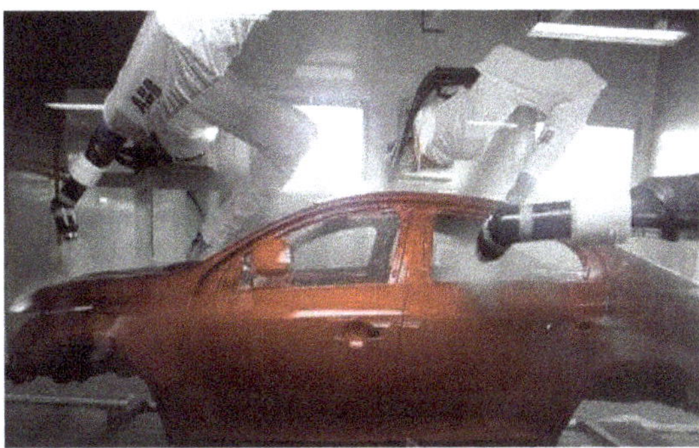

Figure 15. An automobile paint line composed by several spray-painting robots.

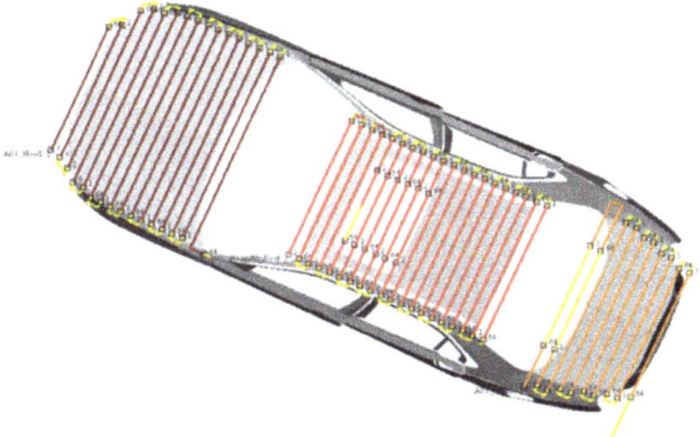

Figure 16. Spray painting trajectory on the roof part of the automobile.

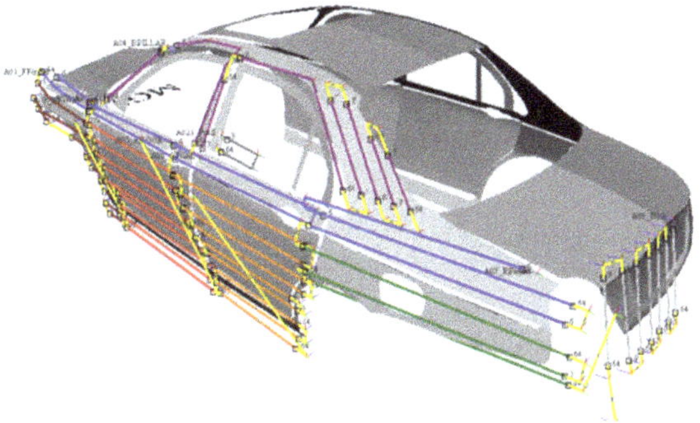

Figure 17. Spray painting trajectory on the left part and the left rear part of the automobile.

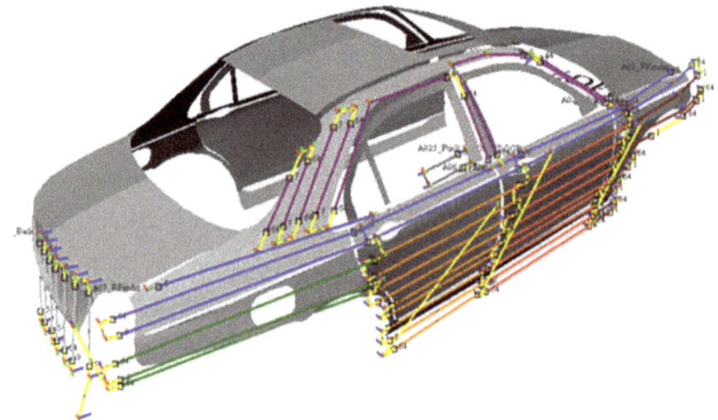

Figure 18. Spray painting trajectory on the right part and the right rear part of the automobile.

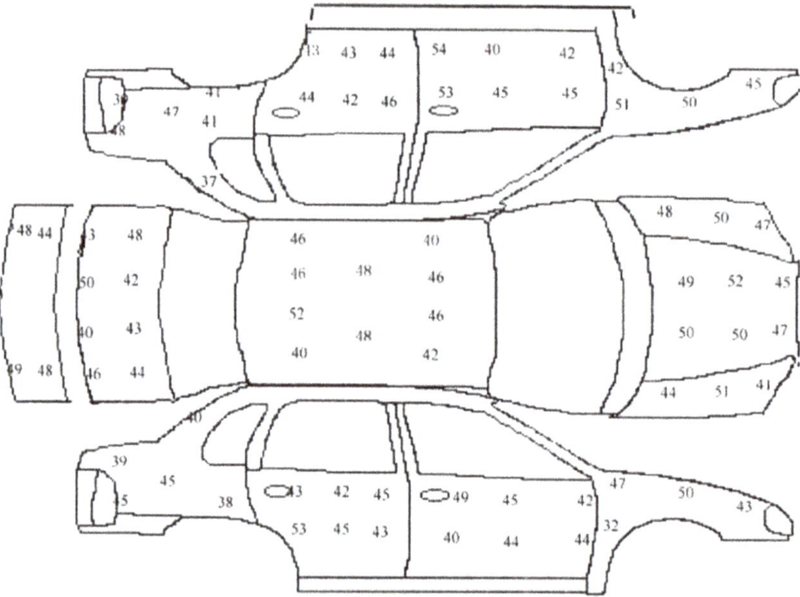

Figure 19. The paint thickness at the sampling points on the automobile body.

From the point of view of spray painting efficiency, the compared results of the GA, ACO, PSO, and random combination are shown in Table 2. It can be seen from the results that the total length of the spray trajectory using the PSO algorithm is the shortest, the spray painting time is the least, and the execution time of the system operation is the longest, which is within the allowable range in the practical application.

Table 2. The compared results of different algorithms.

	GA	ACO	PSO
Total length of Spray Path (m)	124.2	116.9	103.6
Spray painting Time of Robot (s)	282	270	253
Execution Time of Operation (s)	0.9	1.1	1.5

6. Conclusions

In this paper, the spatial trajectory optimization method of a spray painting robot for 3D entity objects was proposed. Firstly, the finite range model of the paint deposition rate was established, and the 3D entity was sliced by the surface modeling method according to FPAG. Then, after planning the spray path on each patch, the variance of the paint thickness of the discrete point and the ideal paint thickness was taken as the objective function and the trajectory on each patch was optimized. The path at the junction of two patches was the PA-PA (parallel-parallel) mode. The improved GA algorithm, ACO algorithm, and PSO algorithm were used to solve the TTOI problem. The practicability of the algorithms was verified by simulation experiments. Finally, spraying experiments were conducted on the off-line programming experimental platform of the spraying robot, and the results of the three algorithms were studied. The results of the automobile body spraying experiments showed that the proposed trajectory optimization of the 3D entity spraying robot can completely satisfy the requirements of uniformity of the spraying thickness. More experimental data please see the Supplementary Materials.

Supplementary Materials: The following are available online at http://www.mdpi.com/2079-9292/8/1/74/s1.

Author Contributions: W.C. and Y.T. conceived and designed the experiments; W.C. performed the experiments; H.L. and J.L. analyzed the data; X.W. contributed reagents/materials/ analyzed the data; X.W. contributed reagents/materials/analysis tools; X.W. wrote the paper.

Funding: This research was supported by the National Natural Science Foundation of China under grant 61503162, "Six talent peaks" high-level talents projection Jiangsu Province under grant 2016GDZB021, Project funded by China Postdoctoral Science Foundation under grant 2016M601691, Industrial Foresight Project of Zhenjiang City under grant GY2018018.

Conflicts of Interest: The authors declare no conflict of interest.

References

1. Gao, W.; He, J. Application of spraying technology on the inner surface of automobile body. *Electroplat. Finish.* **2016**, *22*, 1194–1199.
2. Yuan, A.; Shen, S. Research on Motion Simulation of Spraying Robot Based on ADAMS and MATLAB. *Modular Mach. Tool Autom. Mach. Technol.* **2014**, *8*, 44–48.
3. Akafuah, N.; Poozesh, S.; Salaimeh, A. Evolution of the automotive body coating process—A review. *Coatings* **2016**, *6*, 24. [CrossRef]
4. Akafuah, N.K.; Salazar, A.J.; Saito, K. Infrared thermography-based visualization of droplet transport in liquid sprays. *Infrared Phys. Technol.* **2010**, *53*, 218–226. [CrossRef]
5. Kolakowska, E.; Smith, S.F.; Kristiansen, M. Constraint optimization model of a scheduling problem for a robotic arm in automatic systems. *Robot. Auton. Syst.* **2014**, *62*, 267–280. [CrossRef]
6. Gasparetto, A.; Vidoni, R.; Pillan, D.; Saccavini, E. Automatic path and trajectory planning for robotic spray painting. In Proceedings of the 7th German Conference on Robotics, Munich, German, 21–22 May 2012; pp. 211–216.
7. Atkar, P.N.; Greenfield, A.; Conner, D.C. Uniform coverage of automotive surface patches. *Int. J. Robot. Res.* **2005**, *24*, 883–898. [CrossRef]
8. Posa, M.; Cantu, C.; Tedrake, R. A direct method for trajectory optimization of rigid bodies through contact. *Int. J. Robot. Res.* **2014**, *33*, 69–81. [CrossRef]
9. Yuan, Z. Trajectory planning of Bezier curve based on improved genetic algorithm. *J. Shanghai Dian Ji Univ.* **2012**, *15*, 237–240.
10. From, P.J.; Gunnar, J.; Gravdahl, J.T. Optimal paint gun orientation in spray paint applications—Experimental results. *IEEE Trans. Autom. Sci. Eng.* **2011**, *8*, 438–442. [CrossRef]
11. Jiao, J.; Cao, Z.; Zhao, P.; Liu, X.; Tan, M. Bezier curve based path planning for a mobile manipulator in unknown environments. In Proceedings of the IEEE International Conference on Robotics and Biomimetics, Shenzhen, China, 12–14 December 2013; pp. 1864–1868.
12. Chen, W.; Zhao, D.; Liang, Z. Design of Tool Path Planning of Robotic Spray Painting and Its Experiment. *China Mech. Eng.* **2011**, *17*, 2104–2108.

13. Toljic, N.; Adamiak, K.; Castle, G.S.P. Three-dimensional numerical studies on the effect of the particle charge to mass ratio distribution in the electrostatic coating process. *J. Electrost.* **2014**, *69*, 189–194. [CrossRef]
14. Chen, H.P.; Sheng, W.H. Transformative industrial robot programming in surface manufacturing. In Proceedings of the IEEE International Conference on Robots and Automation, Shanghai, China, 9–13 May 2011; pp. 6059–6064.
15. Chen, H.P.; Xi, N. Automated robot trajectory connection for spray forming process. *J. Manuf. Sci. Eng.* **2012**, *134*, 171–179. [CrossRef]
16. Zeng, Y.; Gong, J.; Xu, N. Tool trajectory optimization of spray painting robot for many-times spray painting. *Int. J. Control Autom.* **2014**, *7*, 193–208. [CrossRef]
17. Wei, C.; Yang, T.; Qiang, Z. A novel trajectory planning scheme for spray painting robot. In Proceedings of the 28th Chinese Control and Decision Conference, Yinchuan, China, 28–30 May 2016; pp. 7008–7012.
18. Chen, W.; Liu, H.; Tang, Y.; Liu, J. Trajectory Optimization of Electrostatic Spray Painting Robots on Curved Surface. *Coatings* **2017**, *7*, 155. [CrossRef]
19. Chen, W.; Liu, J.; Huan, J.; Liu, H. Trajectory Optimization of Spray Painting Robot for Complex Curved Surface Based on Exponential Mean Bézier Method. *Math. Probl. Eng.* **2017**, 4259869. [CrossRef]
20. Li, J. Distributed cooperative tracking of multi-agent systems with actuator fault. *Trans. Inst. Meas. Control* **2015**, *37*, 1041–1048. [CrossRef]
21. Li, J.; Li, C.; Yang, X.; Chen, W. Event-Triggered Containment Control of Multi-Agent Systems with High-Order Dynamics and Input Delay. *Electronics* **2018**, *7*, 343. [CrossRef]

Article

Robust Visual Compass Using Hybrid Features for Indoor Environments

Ruibin Guo [1,*], Keju Peng [1], Dongxiang Zhou [1] and Yunhui Liu [2]

[1] College of Electronic Science, National University of Defense Technology, Changsha 410073, China; keju009@nudt.edu.cn (K.P.); dxzhou@nudt.edu.cn (D.Z.)
[2] Department of Mechanical and Automation Engineering, Chinese University of Hong Kong, Hong Kong 999077, China; yunhui.liu@gmail.com
* Correspondence: guoruibin64@126.com; Tel.: +86-731-845-03239

Received: 15 December 2018; Accepted: 7 February 2019; Published: 16 February 2019

Abstract: Orientation estimation is a crucial part of robotics tasks such as motion control, autonomous navigation, and 3D mapping. In this paper, we propose a robust visual-based method to estimate robots' drift-free orientation with RGB-D cameras. First, we detect and track hybrid features (i.e., plane, line, and point) from color and depth images, which provides reliable constraints even in uncharacteristic environments with low texture or no consistent lines. Then, we construct a cost function based on these features and, by minimizing this function, we obtain the accurate rotation matrix of each captured frame with respect to its reference keyframe. Furthermore, we present a vanishing direction-estimation method to extract the Manhattan World (MW) axes; by aligning the current MW axes with the global MW axes, we refine the aforementioned rotation matrix of each keyframe and achieve drift-free orientation. Experiments on public RGB-D datasets demonstrate the robustness and accuracy of the proposed algorithm for orientation estimation. In addition, we have applied our proposed visual compass to pose estimation, and the evaluation on public sequences shows improved accuracy.

Keywords: visual compass; orientation estimation; hybrid features; plane tracking; vanishing direction; Manhattan World; RGB-D camera

---

1. Introduction

Robust orientation estimation is of great significance in robotics tasks such as motion control, autonomous navigation, and 3D mapping. Orientation can be obtained by utilizing carried sensors like the wheel encoder, inertial measurement unit (IMU) [1–4], or cameras [5–7]. Among these solutions, the visual-based method [8–11] is effective, as cameras can conveniently capture informative images to estimate orientation and position. In the past decades, many simultaneous localization and mapping (SLAM) systems [12,13] and visual odometry (VO) methods [14,15] have been proposed. Payá et al. [5] proposed a global description method based on Radon Transform to estimate robots' position and orientation with the equipped catadioptric vision sensor. These methods show good performance in estimating orientation from captured images. However, local and global maps' construction or loop detection is needed in these approaches to reduce drift error.

For most indoor environments, there exist many parallel and orthogonal lines and planes (called the Manhattan World (MW) [16]). These structural regularities are exploited in studies to estimate drift-free rotation without previous complex techniques (map reconstruction and loop closure) [17–19]. Since 3D geometric structures can easily be calculated by using the camera that provides both depth information and color image with 3 channels (red, green, and blue), called the RGB-D camera. The RGB-D camera has become a popular alternative to monocular and stereo cameras for the purpose of rotation estimation, and estimation accuracy has been prominently improved by using the MW assumption with a RGB-D camera [20–23]. However, a major disadvantage of these MW-based methods is that the number of lines and planes used for tracking the MW axes must be no less than 2, which is the minimal sampling for 3 degrees of freedom (DoF). In practice, robots often encounter harsh environments without lines, and only one plane can be visible, resulting in failure in tracking MW axes to estimate the camera orientation. To address these issues, we select some frames as keyframes and exploit hybrid features (i.e., plane, line, and point) to compute the rotation matrix of each captured frame with respect to its reference keyframe instead of directly aligning with the global MW axes.

In this paper, we propose a robust and accurate approach for orientation estimation using RGB-D cameras. We detected and tracked the normal vectors of multiple planes from depth images, and we detected and matched the line and point features from color images. Then, by utilizing these hybrid features (i.e., plane, line, and point), we constructed a cost function to solve the rotation matrix of each captured frame. Meanwhile, we selected keyframes to reduce drift error and avoid directly aligning each frame with the global MW. Furthermore, we extracted the MW axes based on the normal vectors of orthogonal planes and the vanishing directions of parallel lines, and by aligning the current MW axes with the global MW axes, we refined the aforementioned rotation matrix of each keyframe and achieved drift-free orientation. Experiments showed that our proposed method produces lower drift error in a variety of indoor sequences compared to other state-of-the-art methods.

Our algorithm exploits hybrid features and adds a refinement step for keyframes, which can provide robust and accurate rotation estimation, even in harsh environments, as well as general indoor environments. The contributions of this work are as follows:

- We exploited the hybrid features (i.e., plane, line, and point), which provides reliable constraints in solving the rotation matrix for the majority of indoor environments.
- We refined keyframes' rotation matrix by aligning the current extracted MW axes with the global MW axes, which achieves drift-free orientation estimation.
- We evaluated our proposed approach on the ICL-NUIM and TUM RGB-D datasets, which showed robust and accurate performance.

2. Related Work

Pose estimation obtained by VO or V-SLAM systems has been extensively studied for the purpose of meaningful applications, such as autonomous robots and augmented reality. Rotation estimation is usually considered as a subproblem of pose estimation that consists of rotational and translation components. It has been gradually recognized by researchers that the main source of VO drift is inaccurate rotation estimation [19,20]. In the following discussion, we focus on rotation-estimation methods that exploit structural regularities with RGB-D cameras. These methods utilize surface normals, vanishing points (vanishing directions), or mixed constraints to compute camera orientation.

Surface-normal vectors were exploited to estimate camera orientation because their distribution on the unit sphere (or called Gaussian sphere) is regular and more likely around plane-normal vectors in the current environment, as shown in Figure 1. The work of Straub et al. [21] introduced the Manhattan-Frame model in the surface-normal space and proposed a real-time maximum a posteriori (MAP) inference algorithm to estimate drift-free orientation. Zhou et al. [23] developed a mean-shift paradigm to extract and track planar modes in surface-normal vector distribution on the unit sphere, and achieved drift-free behavior by registering the bundle of planar modes. In the work of Kim et al. [24], orthogonal planar structures were exploited and tracked with an efficient SO(3)-constrained mean-shift algorithm to estimate drift-free rotation. These surface-normal-based methods can provide stable and accurate rotation estimation if the number of observed orthogonal planes is not less than two.

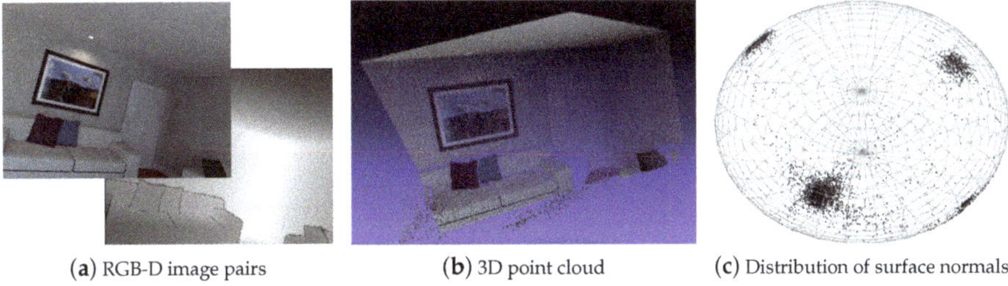

Figure 1. Single RGB-D frame and distribution of surface normals corresponding to its 3D point cloud. (**a**) Frame 85 in 'LivingRoom0' sequence of ICL-NUIM dataset. (**b**) 3D point cloud obtained by back-projecting the depth information and coloring with aligned RGB pixels. (**c**) Distribution of surface normals on the unit sphere.

A vanishing point (VP) of a line is obtained by intersecting the image plane with a ray parallel to the world line and passing through the camera center, and it depends only on the direction of a line [25]. Two parallel lines determine a vanishing direction (VD), and the Euclidean 3D transformation of a VD is influenced only by rotation; the geometric relationships are shown in Figure 2, so VPs and VDs have been widely used for estimating rotation. Bazin et al. [17] proposed a three-line random-sample consensus (RANSAC) algorithm with the VP orthogonality constraint to estimate rotation. The work of Elloumi et al. [26] proposed a real-time pipeline for estimating camera orientation based on vanishing points for indoor navigation assistance on a smartphone. VP-based methods need a sufficient number of lines for estimating rotation, and accuracy performance is greatly affected by line-segment noise.

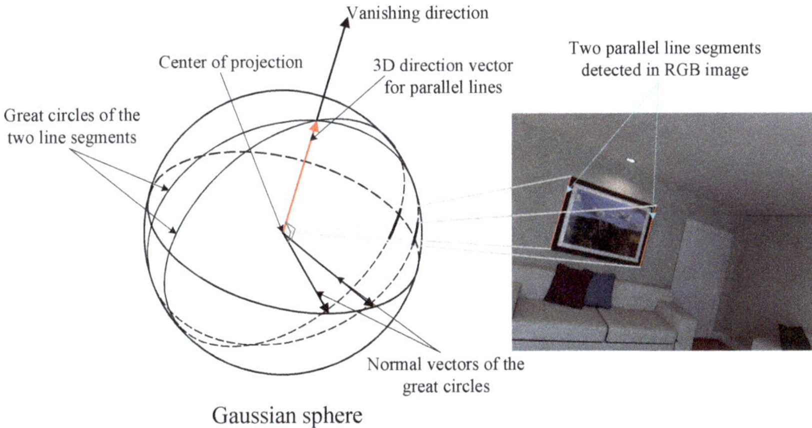

Figure 2. Three-dimensional geometric relationship between parallel lines and their vanishing direction. Gaussian sphere is a unit sphere on the center of a camera projection. Two parallel lines are projected onto the Gaussian sphere as two great circles, and vanishing direction is obtained by the cross project of these two great circles' normal vectors. Two parallel lines and their corresponding vanishing direction are drawn with red.

Hybrid approaches use both surface normals obtained in depth image and vanishing directions extracted in the RGB image to estimate rotation, which shows more robust performance. The method proposed by Kim et al. [22] exploited both line and plane primitives to deal with degenerate cases in surface-normal-based methods for stable and accurate zero-drift rotation estimation. In the work of Kim et al. [27], only a single line and a single plane in RANSAC were used to estimate camera orientation, and refinement is performed by minimizing the average orthogonal distance from the endpoints of the lines parallel to the MW axes once the initial rotation estimation is found. Bazin et al. [17] introduced a related one-line RANSAC for situations where the horizon plane is known.

3. Proposed Method

We propose a robust visual compass that exploits hybrid features (i.e., plane, line, and point) to estimate camera orientation with the RGB and depth-image pairs. Our proposed method has two main steps: (1) rotation matrix is estimated by tracking the hybrid features for each frame with respect to the reference keyframe (tracking step); and (2) refine the keyframe's initial rotation matrix by aligning with the global MW axes to achieve drift-free orientation (refinement step). The overview of our proposed method is shown in Figure 3.

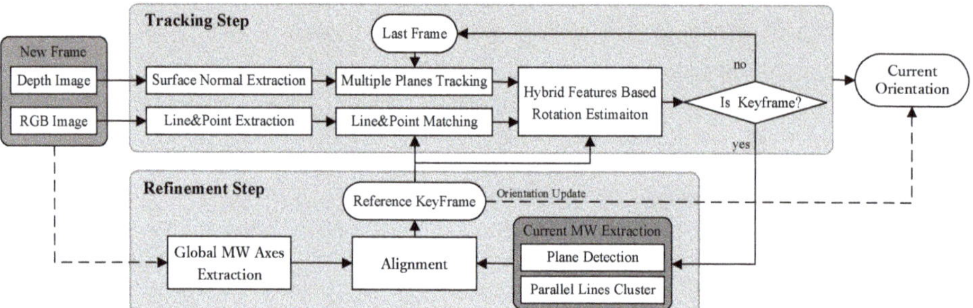

Figure 3. Overview of our visual compass. We estimate camera rotation by tracking the plane, line, and point features. We refine the keyframe orientation by aligning the current Manhattan World (MW) axes with the global MW axes. Global MW axes is extracted from the first captured RGB-D frame.

3.1. Rotation Estimation with Hybrid Features

We simultaneously tracked multiple planes, lines, and points in the current environment, and we utilized tracked hybrid features to construct a cost function for estimating the current rotation matrix relative to its reference keyframe. This can provide camera rotation even in uncharacteristic scenes where there are no rich texture or no consistent visible lines.

3.1.1. Multiple-Plane Detection and Tracking

We detected multiple planes from the depth image with a fast plane extraction algorithm [28]. The algorithm first constructs an initial graph that uniformly divides the depth image's point cloud into a set of nodes with size $H \times W$ in the image space. It then performs agglomerative hierarchical clustering on this graph to merge nodes belonging to the same plane. It final refines the extracted planes using pixel-wise region growing. With this approach, we obtain planes $P_i : (\mathbf{n}_i, d_i), i = 1, ..., m$, where $\mathbf{n}_i$ is the unit normal vector of the i-th plane, and d_i is the distance to the origin of the camera co-ordinate system for the current frame.

We tracked the normal vector of each detected plane with a mean shift algorithm [23] that operates based on the density distribution of initial nodes' normals on the Gaussian sphere, as shown in Figure 4. We calculated the normal vector of each initial node by the least-square fitting method, and the depth image was preprocessed by a box filter to obtain the stable vectors. We used the previous frame's tracked (detected) normal vectors as an initial value, and then performed the mean shift algorithm in the tangent plane of the Gaussian sphere to obtain the tracked results. It should be noted that the parallel planes have the same plane normal vector; we class them as the same plane cluster for rotation estimation.

If we only have plane primitives to estimate rotation, rotation matrix $\boldsymbol{R}$ with three degrees of freedom can be computed from no less than two such tracked norm vectors that are not parallel because each normal vector represents two independent constraints on $\boldsymbol{R}$.

$$\boldsymbol{R} = argmin \sum_{i=1,\ldots,m} \left\| \boldsymbol{R} \cdot \mathbf{n}_i^{ref} - \mathbf{n}_i^k \right\|_2^2 \tag{1}$$

where $\mathbf{n}_i^{ref}$ represents the i-th detected plane normal vector in the reference keyframe, and $\mathbf{n}_i^k$ represents the tracked result of the i-th plane in frame k.

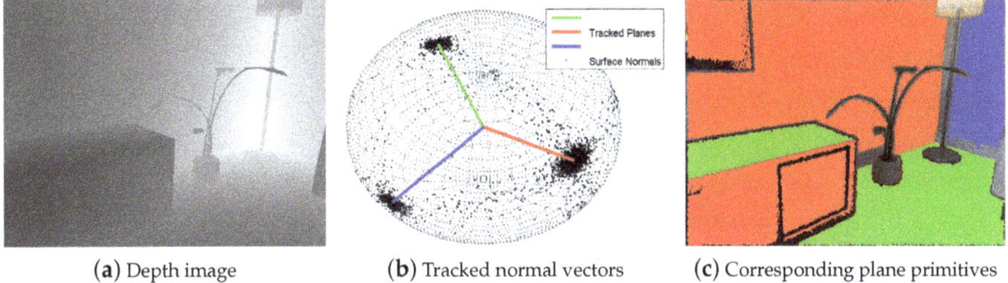

(**a**) Depth image (**b**) Tracked normal vectors (**c**) Corresponding plane primitives

Figure 4. Result of tracking planes: (**a**) Depth image of Frame 741 in the 'LivingRoom2' sequence of the ICL NUIM dataset. (**b**) Tracked normal vectors from the Gaussian sphere; black dots represent the normals of initial nodes. (**c**) Plane primitives in current frame. Tacked normal vectors and their corresponding planes in image domain are represented with the same color.

3.1.2. Line and Point Detection and Matching

We used a linear-time Line Segment Detector (LSD) [29] to extract 2D line segments on the color image and obtain 2D line segments set $\mathbf{l} = \{l_i, i = 1, 2, ..., n\}$, where l_i is the i-th line segment: $y = k_i x + b_i$, with starting point $\mathbf{u}_{si} = (u_{si}, v_{si})$ and ending point $\mathbf{u}_{ei} = (u_{ei}, v_{ei})$. Pixels belonging to the line segment l_i are: $\Re = \{\mathbf{u} | \mathbf{u} \in l_i \wedge \mathbf{u} \in \Omega\}$, Ω is the image domain. Then, we reconstructed their corresponding 3D points set: $\mathbf{P}_l = \{\mathbf{p} | \mathbf{p} = \pi^{-1}(\mathbf{u}, d(\mathbf{u})), \mathbf{p} \in R^3 \wedge \mathbf{u} \in \Re\}$, where $d(\mathbf{u})$ represents the corresponding depth value of pixel $\mathbf{u}$ in the color domain, and $\pi^{-1}(\mathbf{u}, d(\mathbf{u})) = d(\mathbf{u})(\frac{u-c_x}{f_x}, \frac{v-c_y}{f_y}, 1)^T$ is the inverse projection function for a camera model, with f_x,f_y being the focal lengths on the x axis and y axis, and $(c_x, c_y)^T$ is the camera's centre co-ordinates. Finally, the RANSAC method is used to fit 3D lines $\boldsymbol{L}_i : (\mathbf{d}_i, \mathbf{p}_i)$, where $\mathbf{d}_i$ represents the i-th line's 3D direction, and $\mathbf{p}_i$ represents a point in this 3D line.

We match the lines that were respectively extracted from the current frame and the reference keyframe based on the Line Band Descriptor (LBD) [30], and two pairs of matching lines that are not parallel are needed to estimate the rotation matrix in case that there are only line primitives obtained in the current environment.

$$\boldsymbol{R} = argmin \sum_{i=1\ldots n} \left\| \boldsymbol{R} \cdot \mathbf{d}_i^{ref} - \mathbf{d}_i^k \right\|_2^2 \tag{2}$$

where $\mathbf{d}_i^{ref}$ represents the i-th detected 3D line direction in the reference keyframe, and $\mathbf{d}_i^k$ represents the matching line direction in frame k.

In addition to the plane and line features, we extracted and matched the oriented fast and rotated brief (ORB) features for point tracking, as these features are extremely fast to compute and match, and they present good invariance to camera autogain, autoexposure, and illumination changes. We used the epipolar constraint method to optimize the initial matching pairs by ORB descriptors; the optimized results can provide reliable constraints to estimate the camera pose.

$$R = argmin \sum_{i...N} \left\| (R \cdot \pi^{-1}(\mathbf{u}_i^{ref}, d(\mathbf{u}_i^{ref})) + \mathbf{t}) - \pi^{-1}(\mathbf{u}_i^k, d(\mathbf{u}_i^k)) \right\|_2^2 \quad (3)$$

where $\mathbf{u}_i^{ref}$ represents the 2D position of i-th detected ORB point feature in the reference keyframe and $\mathbf{u}_i^k$ represents the 2D position of the matching point in frame k. It should be noted that translation component $\mathbf{t}$ could be obtained by solving Equation (3), but we did not consider it, as our visual compass mainly focused on camera orientation.

3.1.3. Robust Rotation Estimation

We jointly utilized the tracked planes, lines, and points in the current environment to estimate the rotation matrix with respect to the reference keyframe. Rotation matrix R can be computed by solving:

$$R = argmin(\sum_{i=1...m} \lambda_i^P \left\| R \cdot \mathbf{n}_i^{ref} - \mathbf{n}_i^k \right\|_2^2 + \sum_{i=1...n} \lambda_i^L \left\| R \cdot \mathbf{d}_i^{ref} - \mathbf{d}_i^k \right\|_2^2 + \sum_{i=1...N} \left\| (R \cdot P_i^{ref} + \mathbf{t}) - P_i^k \right\|_2^2) \quad (4)$$

where λ_i^P represents the number of pixels contained in the i-th tracked plane, and λ_i^L represents the number of pixels contained in the i-th tracked line, 3D points $P_i^{ref} = \pi^{-1}(\mathbf{u}_i^{ref}, d(\mathbf{u}_i^{ref}))$, and $P_i^k = \pi^{-1}(\mathbf{u}_i^k, d(\mathbf{u}_i^k))$.

Cost function Equation (4) contains three parts, corresponding to plane, line, and point constraints. A stable and accurate rotation matrix can be solved by minimizing Equation (4) with line and plane constraints jointly in the texture-less environment that few points tracked, and the point constraints ensure that the rotation estimation is reliable in the scenes that no consistent lines or only one plane to be visible.

Keyframe Selection: By the tracking step, we constantly know the number of the tracked planes, lines and points for each frame. If there is only one tracked plane with the condition that the number of normal vectors on the Gaussian sphere around this tracked normal vector is too low, and the number of tracked points is less than a threshold, we reuse the fast plane extraction method and Line Segment Detector (LSD) method to detect planes and lines in the current frame. If the number of redetected orthogonal planes and lines is larger than 2, this frame is selected as a keyframe and performs the following refinement step.

3.2. Drift-Free Orientation Estimation

The previous tracking step estimates the rotation matrix between the current frame and its reference keyframe, and it is obvious that the accuracy of the reference keyframe's orientation directly affects the accuracy of the current frame's rotation matrix. To reduce drift error, we sought global MW axes in the first frame and refined each keyframe's orientation by aligning the current extracted MW axes with the global MW axes to achieve drift-free rotation in MW scenes. We sought current and global MW axes based on the plane normal vectors, and the vanishing directions of the parallel lines. Plane normal vectors can be directly obtained by the previous fast plane extraction method, and we propose a novel vanishing direction extraction method as follows.

3.2.1. Vanishing Direction Extraction

To extract accurate VDs, we need to cluster lines that are parallel in the real world. We used the simplified Expectation–Maximization (EM) clustering method to group image lines and compute their corresponding 3D direction vectors. The original EM algorithm iterates the expectation and the maximization steps. In our simplified algorithm, we skip the expectation phase and roughly cluster the lines based on the K-means method [31], with the Euclidean distance of all extracted lines' 3D directions that are represented as 3D points. In the maximization phase, direction vectors are estimated by maximizing the objective function:

$$\mathbf{d}_k^v = argmax\prod_i p(\mathbf{d}_k^v|\mathbf{l}_i^{(k)}) \tag{5}$$

where $\mathbf{l}_i^{(k)}$ represents the i-th 2D line segment in the k-th initial classification, $\mathbf{d}_k^v$ represents the VD of the k-th initial classification to be optimized, and the $p(\mathbf{d}_k^v|\mathbf{l}_i^{(k)})$ represents the posterior likelihood of the VD.

Using the Bayes formula, the posterior likelihood of the VD is expressed as:

$$p(\mathbf{d}_k^v|\mathbf{l}_i^{(k)}) = \frac{p(\mathbf{l}_i^{(k)}|\mathbf{d}_k^v)p(\mathbf{d}_k^v)}{p(\mathbf{l}_i^{(k)})} \tag{6}$$

where $p(\mathbf{l}_i^{(k)}|\mathbf{d}_k^v)$ represents the prior probability of the VD, and $p(\mathbf{d}_k^v)$ models the potential knowledge of the VD before we obtain the measurement. If we know nothing, $p(\mathbf{d}_k^v)$ is defined as uniform distribution with a constant value. Therefore, the VD can be obtained by maximizing prior probability:

$$\mathbf{d}_k^v = argmax\prod_i p(\mathbf{d}_k^v|\mathbf{l}_i^{(k)}) = argmax\sum_i \log p(\mathbf{l}_i^{(k)}|\mathbf{d}_k^v) \tag{7}$$

Prior probability $p(\mathbf{l}_i^{(k)}|\mathbf{d}_k^v)$ is defined as:

$$p(\mathbf{l}_i^{(k)}|\mathbf{d}_k^v) = \frac{1}{\sqrt{2\pi\sigma_k^2}}\left(\frac{-(\mathbf{l}_i^{(k)\mathbf{T}}\mathbf{K}\mathbf{d}_k^v)^2}{2\sigma_k^2}\right) \tag{8}$$

where $\mathbf{K}$ represents the internal camera parameters. Equation (8) reflects the fact that the vanishing direction is perpendicular to the plane normal of a great circle that is determined by an image line $\mathbf{l}_i^{(k)}$ and the center of the projection of the camera, as shown in Figure 2.

Maximizing objective function Equation (7) is equivalent to solving a weighted least-squares problem for each $\mathbf{d}_k^v$:

$$\mathbf{d}_k^v = argmin\sum_i \frac{length(\mathbf{l}_i^{(k)})}{max(length(\mathbf{l}^{(k)}))} \cdot (\mathbf{l}_i^{(k)\mathbf{T}}\mathbf{K}\mathbf{d}_k^v)^2 \tag{9}$$

where $length(\mathbf{l}_i^{(k)})$ represents the length of the i-th line, and $max(length(\mathbf{l}^{(k)}))$ represents the maximum line length in rough cluster k. Length coefficient is considered in the term because the longer the lines are, the more reliable they are. By solving Equation (9), we can obtain the initial vanishing direction and the residual for each line. To obtain a more accurate vanishing direction, we discarded lines with a larger residual than a threshold and added additional optimization. Optimized parallel lines are used to estimate the final vanishing directions. Figure 5 shows two results of parallel-line clustering obtained by our proposed method.

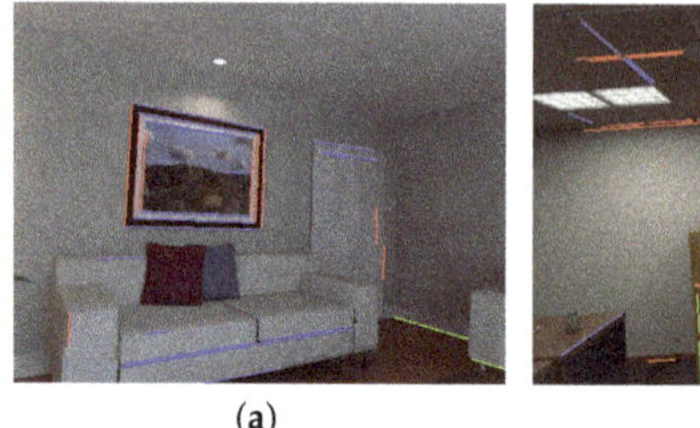
(a)

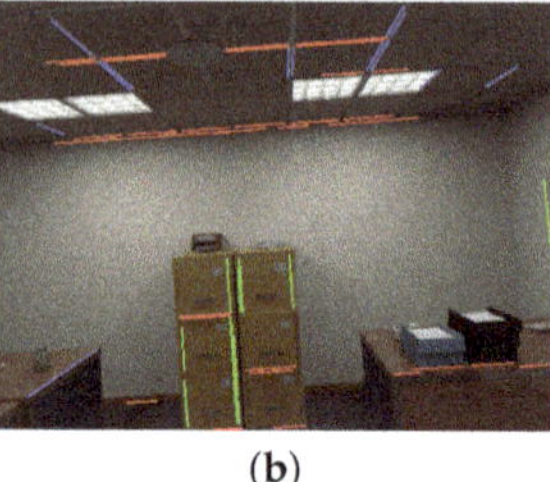
(b)

Figure 5. Results of parallel-line clustering that are used to compute vanishing directions: (**a**) the 120-th image in 'Living Room 2' sequence. (**b**) the 64-th image in 'Office Room 1' sequence. Lines with the same color are parallel in the real world. The vanishing directions obtained by these parallel lines can provide accurate and reliable constraints for MW extraction.

3.2.2. Global Manhattan World Seeking

The Manhattan World assumption is suitable in human-made indoor environments, and we sought global MW axes based on plane normal vectors and the vanishing directions from the first frame. MW axes can be expressed as columns of a 3D rotation matrix $R = [\mathbf{r}_1^g\ \mathbf{r}_2^g\ \mathbf{r}_3^g] \in SO(3)$.

We first set the the detected plane normal vectors and the vanishing directions from the parallel lines as the candidate MW axes, which return a redundant set. In fact, most of the pixels in the frame typically belong to the planes and lines that determine the dominant MW axes, and we sought the plane that contained the largest pixels, and set its plane normal vector $\mathbf{r}_1$ as the first MW axis. The remaining two axes $\mathbf{r}_2$ and $\mathbf{r}_3$ are determined based on the orthogonality constraint with the first axis and the number of the pixels belonging to the detected planes or the parallel lines.

If we detect three mutually orthogonal planes in the first frame, we directly set their plane-normal vectors as the MW axes. In the case that there are only two orthogonal planes, the third axis is determined by the vanishing direction from the parallel lines that is orthogonal with the two previous plane-normal vectors. Similarly, if only one plane is detected in the first frame, we sought the remaining two MW axes by the vanishing directions from the orthogonal parallel lines.

Initial MW axes $[\mathbf{r}_1\ \mathbf{r}_2\ \mathbf{r}_3]$, obtained by the previous step, are not strictly orthogonal. We maintained orthogonality by reprojecting the MW axes onto the closest matrix on $SO(3)$. Each axis is reweighted by a factor λ_i that is determined by the number of pixels belong to this axis' corresponding planes or parallel lines. The final global MW axes are obtained by using singular=value decomposition (SVD):

$$\left[\mathbf{r}_1^g\ \mathbf{r}_2^g\ \mathbf{r}_3^g\right] = \mathbf{U}\mathbf{V}^T \tag{10}$$

where $[\mathbf{U}, \mathbf{D}, \mathbf{V}] = SVD([\lambda_1\mathbf{r}_1\ \ \lambda_2\mathbf{r}_2\ \ \lambda_3\mathbf{r}_3])$ and factor λ_i describes how certain the observation of a direction is.

3.2.3. Keyframe Orientation Refinement

For each keyframe, we refined its rotation matrix by aligning the current extracted MW axes with the global MW axes. We first used the fast plane-extraction algorithm and our proposed spatial-line direction estimation method to extract plane-normal vectors and vanishing directions in the current keyframe. We then extracted the current MW axes $(\mathbf{r}_i^c, i = 1, 2, 3)$ by using the same method that we used to extract global MW axes. We finally determine the corresponding pairs based on the Euclidean distance between vector $\mathbf{r}_i^c$ and $\mathbf{R}\mathbf{r}_j^g$, where $\mathbf{R}$ represents the rotation matrix obtained by tracking steps for the current keyframe. Refined rotation matrix $\mathbf{R}^r$ is computed by solving:

$$\mathbf{R}^r = argmin \sum_{i=1,2,3} \left\| \mathbf{R}^r \mathbf{r}_i^g - \mathbf{r}_i^c \right\|_2^2 \tag{11}$$

where $\mathbf{r}_i^g$ and $\mathbf{r}_i^c$ represent the i-th global MW axis and the current extracted corresponding MW axis.

4. Results

We evaluate our proposed approach on the synthetic dataset (ICL-NUIM [32]), real-world dataset (TUM RGB-D [33]), and pose-estimation application, respectively. All experiments were run on a desktop computer with an Intel Core i7, 16 GB memory, and Ubuntu 16.04 platform.

- The *ICL-NUIM* dataset is a collection of handheld RGB-D camera sequences within synthetically generated environments. These sequences were captured in a living room and an office room with perfect ground-truth poses to fully quantify the accuracy of a given visual odometry or SLAM system. Depth and RGB noise models were used to alter the ground images to simulate realistic sensor noise. There are sequences that are captured in the environment with low texture and only one visible plane, which makes it hard to estimate rotation for whole images in this sequence.
- The *TUM RGB-D* dataset is a famous benchmark to evaluate the accuracy of a given visual odometry or visual SLAM system. It contains various indoor sequences captured from the Kinect RGB-D sensor. The sequences were recorded in real environments at a frame rate of 30 Hz with a 640×480 resolution, and their ground-truth trajectories were obtained from a high-accuracy motion-capture system. The TUM dataset is more challenging than the ICL dataset because it has some blurred images and inaccurate alignment image pairs that make it difficult to estimate the rotation matrix.

We compared our proposed approach with two state of the art MW-based methods proposed by Zhou et al. [23] and Kim et al. [27], namely, orthogonal planes based rotation estimation (OPRE) and 1P1L. OPRE estimates absolute and drift-free rotation by exploiting orthogonal planes from depth images. 1P1L estimates 3DoF drift-free rotational motion with only a single line and plane in the Manhattan world. We used the average value of the absolute rotation error (ARE) in degrees as the performance metric for the entire sequences:

$$\mathbf{ARE}.average = \frac{1}{N}\sum_{i=1}^{N} \arccos(\frac{tr(\mathbf{R}_i^T \cdot \mathbf{R}_i^g) - 1}{2}) \times 57.3 \tag{12}$$

where $tr()$ denotes the trace of a matrix, $\mathbf{R}_i$ and $\mathbf{R}_i^g$ represent the estimated and ground rotation matrix for the i-th frame, respectively, and N represents the number of frames in the tested sequence.

4.1. Evaluation on Synthetic Dataset

We first tested the performance of our proposed algorithm on the ICL-NUIM dataset, and measured the average ARE in degrees for each sequence; evaluation results are shown in Table 1. The smallest average ARE values are bolded, which reveals that our proposed method is more accurate than the two other methods. For example, in 'Office Room 0', the average ARE of our proposed method is 0.16 degrees, while that of 1P1L and OPRE is 0.37 and 0.18 degrees, respectively. Our method outperformed the others in all cases in the ICL-NUIM benchmark. The main reason is that we jointly exploited the plane, line, and point features to estimate camera orientation even when the camera moves in scenes with no consistent lines, or where only one plane is visible; this is illustrated in Figure 6. In 'Living Room 0', the OPRE method failed to estimate the rotation for the entire sequence because only one plane can be visible in some frames; we marked the result as '$\times$' in Table 1. The last column in Table 1 shows the number of frames in the current sequence.

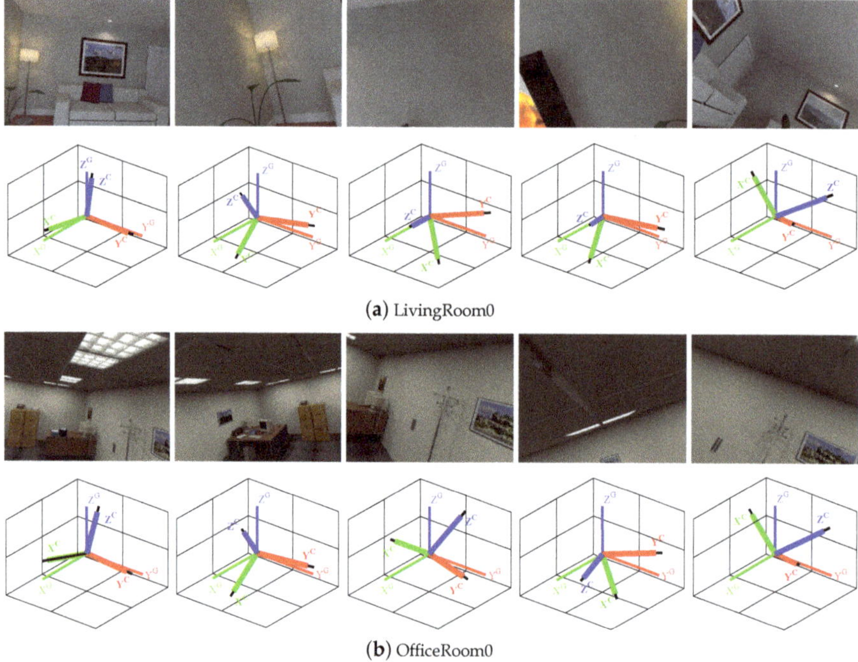

(a) LivingRoom0

(b) OfficeRoom0

Figure 6. Results of camera orientation estimated by our proposed method on the ICL-NUIM dataset: (a) 'Living Room 0' sequence. (b) 'Office Room 0' sequence. The estimated orientation of each frame is shown in the bottom of the RGB image. Colored thick and thin lines respectively denote current orientation and global MW axes (determined in the first frame); black lines represent ground orientation.

Table 1. Comparison of average absolute rotation error (ARE, degrees) on the ICL-NUIM dataset.

Sequence	Proposed	No Refinement	1P1L	OPRE	Frames
Living Room 0	**0.22**	0.53	0.31	×	1508
Living Room 1	**0.25**	0.55	0.38	0.97	965
Living Room 2	**0.23**	1.26	0.34	0.49	880
Living Room 3	**0.35**	1.69	0.35	1.34	1240
Office Room 0	**0.16**	1.32	0.37	0.18	1507
Office Room 1	**0.17**	0.44	0.37	0.32	965
Office Room 2	**0.26**	1.29	0.38	0.33	880
Office Room 3	**0.14**	0.43	0.38	0.21	1240

The MW assumption is sufficiently suitable for the ICL-NUIM benchmark and we used the refinement step to achieve a drift-free rotation matrix. To clearly show the effect of the refinement step, we measured the ARE values in degrees for all sequences by our method without a refinement step, which corresponds to the 'No Refinement' column in Table 1. We recorded the values of absolute rotation error (ARE) for each frame in the 'Living Room 0' sequence, and the final rotation drift with and without refinement was 0.34 and 1.43 degrees, respectively, as shown in Figure 7. This demonstrates that the refinement step can effectively reduce rotation drift.

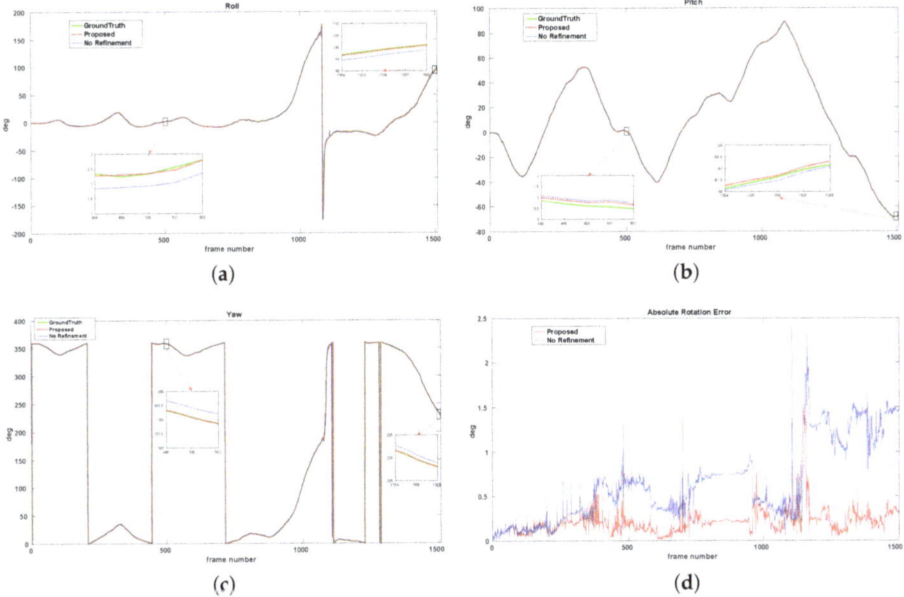

Figure 7. Performance evaluation for refinement on the 'Living Room 0' sequence: (**a**–**c**) Roll, pitch, and yaw angles estimated by the proposed method with and without a refinement step for each frame. (**d**) Absolute rotation errors for our proposed methods with and without a refinement step. This shows that the refinement step can effectively reduce rotation drift.

4.2. Evaluation on Real-World Data

We compared the performance of our proposed algorithm with other two methods on seven real-world TUM RGB-D sequences that contained structural regularities. Comparison results are shown in Table 2. We provide the average ARE for rotation estimation, and the smallest values are indicated in bold. Our proposed method showed better performance in low-texture environments such as 'fr3_struc_notex' and 'fr3_cabinet' because we used hybrid features to estimate orientation, as shown in Figure 8. Our method can also provide a more accurate rotation matrix in an environment with imperfect MW structure like 'fr3_nostruc_tex' and 'fr3_nostruc_notex', whereas OPRE fails because it requires at least two orthogonal planes to estimate camera orientation.

Table 2. Comparison of average ARE (degrees) on TUM RGBD Dataset.

Sequence	Proposed	No Refinement	1P1L	OPRE	Frames
fr3_nostruc_notex	**1.22**	1.22	1.51	×	90
fr3_nostruc_tex	**1.89**	1.89	2.15	×	448
fr3_struc_notex	**1.20**	1.62	1.96	3.01	965
fr3_struc_tex	**0.74**	1.03	2.92	3.81	905
fr3_cabinet	**1.48**	2.78	2.48	2.42	926
fr3_large_cabinet	**1.87**	3.95	2.04	36.34	980
fr3_long_office	**1.51**	3.58	1.75	4.99	2486

The result of refinement performance on the 'fr3_cabinet' sequence is shown in Figure 9. Final rotation drift with and without refinement was 1.30 and 1.62 degrees, respectively. It is clear that our refinement step can effectively reduce drift error. The average ARE values computed by our proposed algorithm with and without refinement step were the same in sequences 'fr3_nostruc_tex' and 'fr3_nostruc_notex'. The reason is that there were no perfect global MW axes extracted in the first frame, and the refinement step was not implemented.

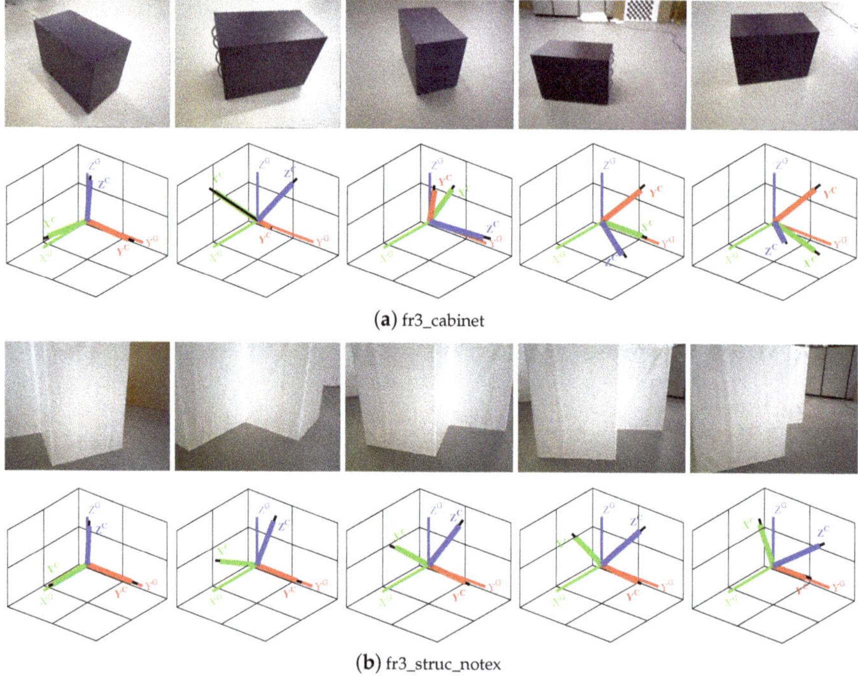

(**a**) fr3_cabinet

(**b**) fr3_struc_notex

Figure 8. Results of camera orientation estimated by our proposed method on the TUM RGB-D dataset: (**a**) 'fr3_cabinet' sequence. (**b**) 'fr3_struc_notex' sequence. Estimated orientation of each frame is shown in the bottom of the RGB image. Colored thick and thin lines respectively denote current orientation and global MW axes (determined in the first frame); black lines represent ground orientation.

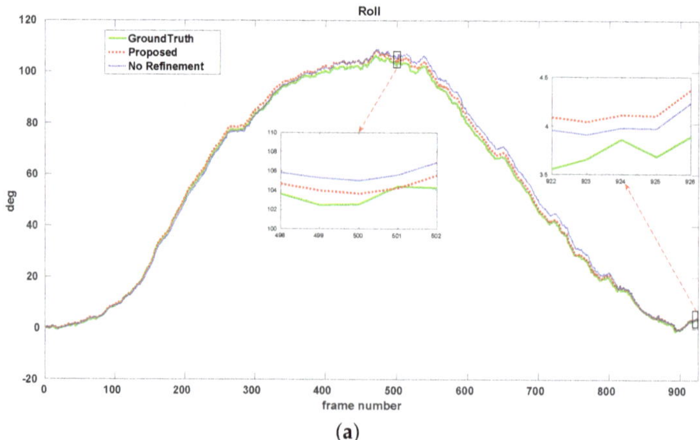

(**a**)

Figure 9. *Cont.*

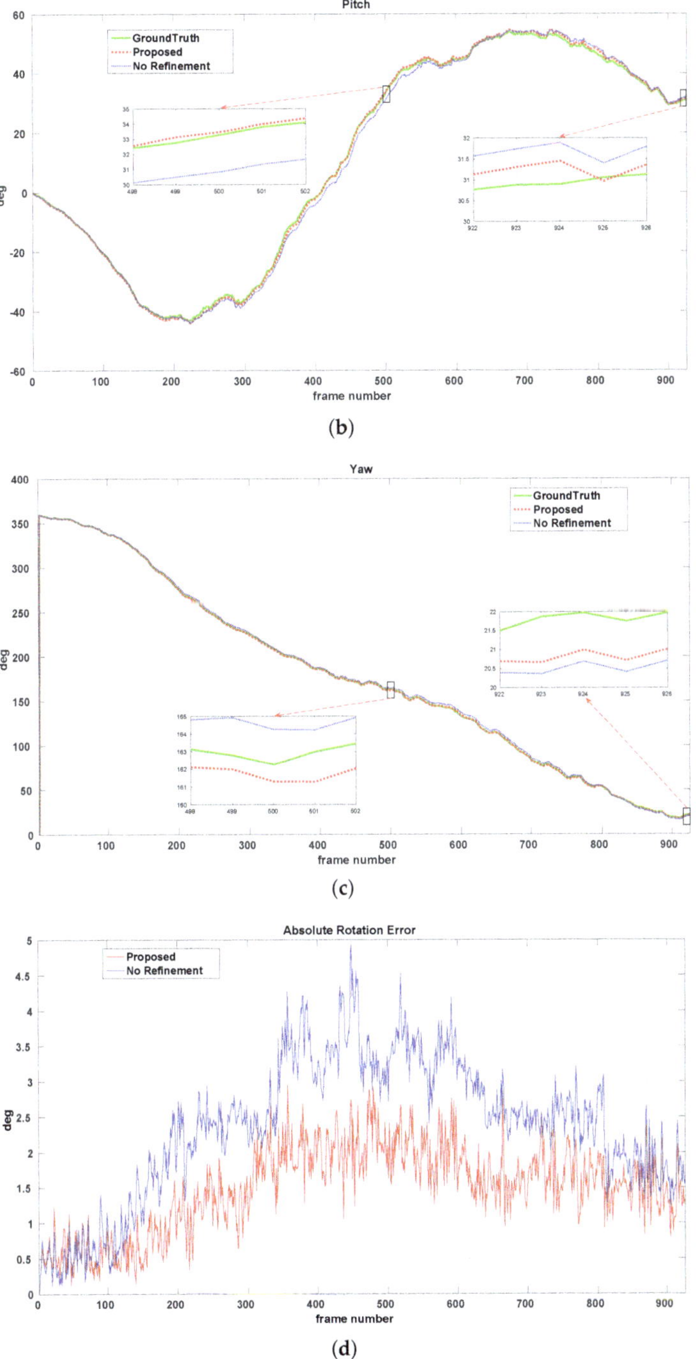

Figure 9. Performance evaluation for refinement on the 'fr3_cabinet' sequence: (**a**–**c**) Roll, pitch, and yaw angles estimated by our proposed method with and without a refinement step for each frame. (**d**) Absolute rotation errors for the proposed methods with and with refinement step. This shows that the refinement step can effectively reduce rotation drift.

4.3. Application to Pose Estimation

To further verify the practicability of our proposed visual compass, we used it for pose-estimation application and recorded the estimated trajectories. The three-dimensional pose has six degrees of freedom (DoF) and it consists of 3-DoF rotation and 3-DoF translation. As our proposed visual compass method can provide accurate rotation estimation, the key to performing pose localization is to estimate the translation component. We first detected and tracked ORB feature points to obtain point correspondences. Then, we recovered the 3-DoF translational motion of the images by minimizing:

$$t = argmin(\sum_{i=1...N} \left\| (\boldsymbol{R}^{fixed} \cdot \boldsymbol{P}_i^{ref} + \mathbf{t}) - \boldsymbol{P}_i^k \right\|_2^2) \tag{13}$$

where $\boldsymbol{R}^{fixed}$ represents the rotation matrix between the reference image and the current image, and it is obtained by our proposed visual compass, Three-dimensional points $\boldsymbol{P}_i^{ref}$ and $\boldsymbol{P}_i^k$ are described in Equation (4).

We tested pose estimation on four datasets, "Living Room 2", "Office Room 3", "fr3_struc_tex", and "fr3_nostruc_tex". These datasets provide the ground-truth pose for each image; we measured the root mean squared error (RMSE) of the absolute translational error (ATE) and compared it with state-of-the-art approaches, namely, ORB_SLAM [12], dense visual odometry (DVO) [15], and line-plane based visual odometry (LPVO) [22]. The comparison of ATE.RMSE is shown in Table 3; the smallest error for each sequence is indicated in bold. Estimated trajectories are shown in Figure 10.

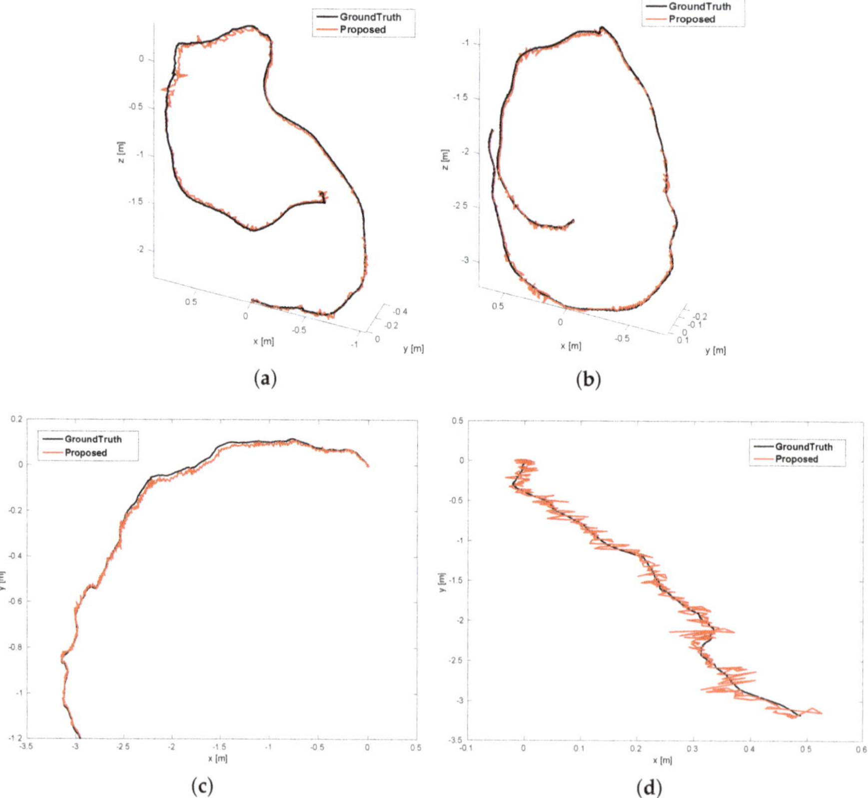

Figure 10. Estimated trajectories with the proposed (red) and ground truth (black) on four sequences: (**a**) Living Room2, (**b**) Office Room 3, (**c**) fr3_struc_tex, (**d**) fr3_nostruc_tex.

Table 3. Comparison of ATE.RMSE (unit: m).

Sequence	Proposed	ORB-SLAM2	DVO	LPVO
Living Room 2	**0.025**	0.028	0.375	0.034
Office Room 3	**0.021**	0.065	0.079	0.030
fr3_struc_tex	**0.017**	0.024	0.048	0.174
fr3_nostruc_tex	**0.046**	0.052	0.073	×

5. Conclusions

We proposed a visual-based method to estimate robot orientation with RGB-D cameras. We exploited hybrid features providing reliable constraints to construct cost function for solving the initial rotation matrix. We presented a vanishing direction extraction method based on parallel lines and combined it with detected plane normals to seek global and current Manhattan World axes. We refined the orientation matrix of the selected keyframe with respect to the global MW axes, and achieved drift-free orientation. The proposed algorithm was tested on both synthetic as well as real-world publicly available RGB-D datasets, and we compared it with two state-of-the-art methods for orientation estimation. The results demonstrated the accuracy and robustness of our proposed method. Furthermore, we applied the proposed algorithm to pose estimation and recovered the translational motion by giving absolute camera orientation; evaluation on the public sequences showed improved accuracy. In summary, the proposed algorithm showed good performance in Manhattan World scenes, and it has significant applications on mobile robotics. In the future, we will exploit hybrid features to perform pose location and 3D mapping that can provide maps with a geometric structure and more robust pose estimation in less-textured and -structured environments, and we will try to optimize the refinement step for not only pure Manhattan Worlds but also more general environments like Mixtures of Manhattan Worlds [34].

Author Contributions: Methodology, R.G.; resources, R.G.; software, R.G.; supervision, K.P. and D.Z.; validation, K.P. and Y.L.; writing—original draft, R.G.; writing—review and editing, D.Z. and Y.L.

Funding: This research received no external funding.

Conflicts of Interest: The authors declare no conflict of interest.

References

1. Reuper, B.; Becker, M.; Leinen, S. Benefits of Multi-Constellation/Multi-Frequency GNSS in a Tightly Coupled GNSS/IMU/Odometry Integration Algorithm. *Sensors* **2018**, *18*, 3052. [CrossRef]
2. Wang, Z.; Chen, Y.; Mei, Y.; Yang, K.; Cai, B. IMU-Assisted 2D SLAM Method for Low-Texture and Dynamic Environments. *Appl. Sci.* **2018**, *8*, 2534. [CrossRef]
3. Fu, Q.; Li, S.; Liu, Y.; Zhou, Q.; Wu, F. Automatic Estimation of Dynamic Lever Arms for a Position and Orientation System. *Sensors* **2018**, *18*, 4230. [CrossRef] [PubMed]
4. Hou, R.; Zhai, L.; Sun, T. Steering Stability Control for a Four Hub-Motor Independent-Drive Electric Vehicle with Varying Adhesion Coefficient. *Energies* **2018**, *11*, 2438. [CrossRef]
5. Payá, L.; Reinoso, O.; Jiménez, L.M.; Juliá, M. Estimating the position and orientation of a mobile robot with respect to a trajectory using omnidirectional imaging and global appearance. *PLoS ONE* **2017**, *12*, e0175938. [CrossRef] [PubMed]
6. Yoon, S.J.; Kim, T. Development of Stereo Visual Odometry Based on Photogrammetric Feature Optimization. *Remote Sens.* **2019**, *11*, 67. [CrossRef]
7. Li, J.; Gao, W.; Li, H.; Tang, F.; Wu, Y. Robust and Efficient CPU-Based RGB-D Scene Reconstruction. *Sensors* **2018**, *18*, 3652. [CrossRef] [PubMed]
8. Perdices, E.; Cañas, J.M. SDVL: Efficient and Accurate Semi-Direct Visual Localization. *Sensors* **2019**, *19*, 302. [CrossRef] [PubMed]
9. Wang, R.; Di, K.; Wan, W.; Wang, Y. Improved Point-Line Feature Based Visual SLAM Method for Indoor Scenes. *Sensors* **2018**, *18*, 3559. [CrossRef] [PubMed]

10. Zhu, J.; Li, Q.; Cao, R.; Sun, K.; Liu, T.; Garibaldi, J.; Li, Q.; Liu, B.; Qiu, G. Indoor Topological Localization Using a Visual Landmark Sequence. *Remote Sens.* **2019**, *11*, 73. [CrossRef]
11. Valiente, D.; Gil, A.; Reinoso, Ó.; Juliá, M.; Holloway, M. Improved omnidirectional odometry for a view-based mapping approach. *Sensors* **2017**, *17*, 325. [CrossRef] [PubMed]
12. Mur-Artal, R.; Tardós, J.D. Orb-slam2: An open-source slam system for monocular, stereo, and rgb-d cameras. *IEEE Trans. Robot.* **2017**, *33*, 1255–1262. [CrossRef]
13. Engel, J.; Schöps, T.; Cremers, D. LSD-SLAM: Large-scale direct monocular SLAM. In *European Conference on Computer Vision*; Springer: Cham, Switzerland, 2014; pp. 834–849.
14. Engel, J.; Koltun, V.; Cremers, D. Direct sparse odometry. *IEEE Trans. Pattern Anal. Mach. Intell.* **2018**, *40*, 611–625. [CrossRef] [PubMed]
15. Kerl, C.; Sturm, J.; Cremers, D. Dense visual SLAM for RGB-D cameras. In Proceedings of the 2013 IEEE/RSJ International Conference on Intelligent Robots and Systems, Tokyo, Japan, 3–7 November 2013; pp. 2100–2106.
16. Coughlan, J.M.; Yuille, A.L. Manhattan world: Compass direction from a single image by bayesian inference. In Proceedings of the Seventh IEEE International Conference on Computer Vision, Kerkyra, Greece, 20–27 September 1999; Volume 2, pp. 941–947.
17. Bazin, J.C.; Pollefeys, M. 3-line RANSAC for orthogonal vanishing point detection. In Proceedings of the 2012 IEEE/RSJ International Conference on Intelligent Robots and Systems, Vilamoura, Portugal, 7–12 October 2012; pp. 4282–4287.
18. Joo, K.; Oh, T.H.; Kim, J.; So Kweon, I. Globally optimal Manhattan frame estimation in real-time. In Proceedings of the IEEE Conference on Computer Vision and Pattern Recognition, Las Vegas, NV, USA, 27–30 June 2016; pp. 1763–1771.
19. Straub, J.; Bhandari, N.; Leonard, J.J.; Fisher, J.W. Real-time manhattan world rotation estimation in 3d. In Proceedings of the 2015 IEEE/RSJ International Conference on Intelligent Robots and Systems (IROS), Hamburg, Germany, 28 September–2 October 2015; pp. 1913–1920.
20. Zhou, Y.; Kneip, L.; Li, H. Real-time rotation estimation for dense depth sensors in piece-wise planar environments. In Proceedings of the 2016 IEEE/RSJ International Conference on Intelligent Robots and Systems (IROS), Daejeon, Korea, 9–14 October 2016; pp. 2271–2278.
21. Straub, J.; Freifeld, O.; Rosman, G.; Leonard, J.J.; Fisher, J.W. The manhattan frame model—Manhattan world inference in the space of surface normals. *IEEE Trans. Pattern Anal. Mach. Intell.* **2018**, *40*, 235–249. [CrossRef] [PubMed]
22. Kim, P.; Coltin, B.; Kim, H.J. Low-drift visual odometry in structured environments by decoupling rotational and translational motion. In Proceedings of the 2018 IEEE International Conference on Robotics and Automation (ICRA), Brisbane, Australia, 21–25 May 2018; pp. 7247–7253.
23. Zhou, Y.; Kneip, L.; Rodriguez, C.; Li, H. Divide and conquer: Efficient density-based tracking of 3D sensors in Manhattan worlds. In *Asian Conference on Computer Vision*; Springer: Cham, Switzerland, 2016; pp. 3–19.
24. Kim, P.; Coltin, B.; Kim, H.J. Visual odometry with drift-free rotation estimation using indoor scene regularities. In Proceedings of the 2017 British Machine Vision Conference, London, UK, 4–9 September 2017.
25. Hartley, R.; Zisserman, A. *Multiple View Geometry in Computer Vision*; Cambridge University Press: Cambridge, UK, 2003.
26. Elloumi, W.; Treuillet, S.; Leconge, R. Real-time camera orientation estimation based on vanishing point tracking under Manhattan World assumption. *J. Real-Time Image Process.* **2017**, *13*, 669–684. [CrossRef]
27. Kim, P.; Coltin, B.; Kim, H.J. Indoor RGB-D Compass from a Single Line and Plane. In Proceedings of the IEEE Conference on Computer Vision and Pattern Recognition, Salt Lake City, UT, USA, 18–22 June 2018; pp. 4673–4680.
28. Feng, C.; Taguchi, Y.; Kamat, V.R. Fast plane extraction in organized point clouds using agglomerative hierarchical clustering. In Proceedings of the 2014 IEEE International Conference on Robotics and Automation (ICRA), Hong Kong, China, 31 May–7 June 2014; pp. 6218–6225.
29. Von Gioi, R.G.; Jakubowicz, J.; Morel, J.M.; Randall, G. LSD: A line segment detector. *Image Process. Line* **2012**, *2*, 35–55. [CrossRef]
30. Zhang, L.; Koch, R. An efficient and robust line segment matching approach based on LBD descriptor and pairwise geometric consistency. *J. Vis. Commun. Image Represent.* **2013**, *24*, 794–805. [CrossRef]

31. Wagstaff, K.; Cardie, C.; Rogers, S.; Schrödl, S. Constrained k-means clustering with background knowledge. *ICML Int.Conf. Mach. Learn.* **2001**, *1*, 577–584.
32. Handa, A.; Whelan, T.; Mcdonald, J.; Davison, A.J. A benchmark for RGB-D visual odometry, 3D reconstruction and SLAM. In Proceedings of the IEEE International Conference on Robotics and Automation, Hong Kong, China, 31 May–7 June 2014; pp. 1524–1531.
33. Sturm, J.; Engelhard, N.; Endres, F.; Burgard, W.; Cremers, D. A benchmark for the evaluation of RGB-D SLAM systems. In Proceedings of the 2012 IEEE/RSJ International Conference on Intelligent Robots and Systems, Vilamoura, Portugal, 7–12 October 2012; pp. 573–580.
34. Straub, J.; Rosman, G.; Freifeld, O.; Leonard, J.J.; Iii, J.W.F. A Mixture of Manhattan Frames: Beyond the Manhattan World. In Proceedings of the Computer Vision and Pattern Recognition, Columbus, OH, USA, 23–28 June 2014; pp. 3770–3777.

Article

Estimated Reaction Force-Based Bilateral Control between 3DOF Master and Hydraulic Slave Manipulators for Dismantlement

Karam Dad Kallu, Jie Wang, Saad Jamshed Abbasi and Min Cheol Lee *

School of Mechanical Engineering, Pusan National University, Busandaehak-ro 63beon-gil, Geumjeong-gu, Busan 46241, Korea; karamdadkallu@gmail.com (K.D.K.); woaiwoffff@gmail.com (J.W.); saadjamshed93@gmail.com (S.J.A.)

* Correspondence: mclee@pusan.ac.kr; Tel.: +82-10-3861-2688

Received: 4 August 2018; Accepted: 15 October 2018; Published: 16 October 2018

Abstract: This paper proposes a novel bilateral control design based on an estimated reaction force without a force sensor for a three-degree of freedom hydraulic servo system with master–slave manipulators. The proposed method is based upon sliding mode control with sliding perturbation observer (SMCSPO) using a bilateral control environment. The sliding perturbation observer (SPO) estimates the reaction force at the end effector and second link without using any sensors. The sliding mode control (SMC) is used as a bilateral controller for the robust position tracking and control of the slave device. A bilateral control strategy in a hydraulic servo system provides robust position and force tracking between master and slave. The difference between the reaction force of the slave produced by the effect of the remote environment and the operating force applied to the master by the operator is expressed in the target impedance model. The impedance model is applied to the master and allows the operator to feel the reaction force from the environment. This research experimentally verifies that the slave device can follow the trajectory of the master device using the proposed bilateral control strategy based on the estimated reaction force. This technique will be convenient for three or more degree of freedom (DOF) hydraulic servo systems used in dismantling nuclear power plants. It is worthy to mention that a camera is used for visual feedback on the safety of the environment and workspace.

Keywords: Hydraulic Servo System; SMCSPO; Bilateral Control; Estimated Reaction Force; Master–Slave Configuration and Nuclear Power Plant

---

1. Introduction

This is an era for computer science and technology; thus, automation and remote operation requires time. The automated/remote system must consider, in detail, the manual operation to come up with an alternative solution that provides better control for the operator with instrumental feedback. Several conditions need to be considered when designing and implementing remote activities. These conditions include a clear workspace for designed equipment, the environmental conditions of the area of operation, handling of different types of materials, tool setup and change out among others. The crux of the story is to replace manual tools with a suitable remote system. A simple example is to modify grip operation by utilizing the end effector and implementing remote alignment and pinning methods and using self-standing bails. A successful remote design always inherits a defined workspace for equipment with minimum mechanical limitations to accomplish a task. Remote technology is a broader area that could be subdivided into different fields such as dismantling equipment, cutting tools, segmenting, sampling and workspace to handle the designed equipment. This field has some state-of-the-art applications, e.g., dismantling of nuclear facilities which require more safety. Several articles in past research [1–8] have been published to address this problem. The Maestro robot system was developed in France to dismantle inactive nuclear reactors [9,10].

An alternative for such activities is tele-operation systems which have human control to ensure safety. In tele-operated systems, human personnel set visual/instrumental feedback from the manipulator to decide future tasks. The controller, in this case, acts according to the master–slave configuration. Bilateral control architecture is at favorable positions for such targets, and most research has presented two-channel architecture such as position–position (P-P), position–force (P-F), force–force (F-F) and force–position (F-P). Several other authors in the past [11–14] have presented bilateral controllers with more than a two-channel architecture.

Several researchers in the past have discussed the improvement in control architecture and the performance of controllers [15,16]. Yana et al. [17] have addressed the finite-time control problem for the bilateral tele-operated environment through resulting feedback. They proposed an observer to estimate a velocity profile by ensuring semi-global stability depending upon the resulting velocity error. A promising feature of this method is that it only utilizes position information which forces the master–slave synchronization error to approach null value in a defined time. Dinh et al. [18] have presented the idea of utilizing a joystick controller to be used for construction machinery control. The authors proposed a controller that is comprised of a force-reflecting gain tuner and a couple of adaptive controllers, namely, master and slave. They implemented a fuzzy logic technique to design local adaptive controllers. Additionally, they utilized an efficient optimization algorithm that provides a real feeling of interaction for an operator at a remote site. Ollin et al. [19] introduced the idea of communications delayed under-actuated mechanical systems under master–slave tele-operated control. They proposed a solution resulting in a particular compiling matrix that ensures the error dynamics refrained in the linear and non-linear parts inherit the matching condition to be satisfied. Later, these linear and non-linear parts are utilized to design discontinuous casual controllers to achieve bilateral co-ordination in position and time. Rabah et al. [20] proposed the idea of using a two adaptive fuzzy controller. The proposed algorithm not only adjusts membership function by the compensatory fuzzy controller but also implements a compensatory learning algorithm for optimal solutions. The first controller is designed under a compensatory neural–fuzzy interface whereas the other is designed under a compensatory adaptive neural–fuzzy interface system. The force–position scheme is utilized by incorporating a two-channel bilateral tele-operated architecture. Finally, the stability and transparency analysis is carried out under passivity framework [20]. Another design of a bilateral controller is presented in Umar et al. [21]. The design is based upon the state of convergence and was implemented for tele-operated systems. The authors applied Takagi–Sugeno's (TS) fuzzy model approach for approximation. The authors mentioned that SC-based bilateral controllers have several advantages in modeling and control design. The most advantageous feature is the ease to implement. The master–slave system in the modeling step can be performed by an *n-th* order differential equation. Similarly, its control design step can easily identify the gains required for the desired closed loop dynamics for a particular system.

Another novel control strategy is presented in [22], where the controller is not placed with the system under experiment. The authors analyzed the stability and transparency of the proposed tele-operators. They implemented PD-like controllers for fixed-time delays and P-like controllers for time-varying delays. Another interesting study by Kamran et al. [23], which utilizes a robust controller for master control and an adaptive back-stepping controller is designed for the slave controller. The authors analyzed the scenarios of input-delay uncertainties in the parameters and multi-objective optimization in implementing the robust master control. KD et al. [24] presented the estimating methodology of reaction force for assembly work with a robot manipulator consisting of a three-link dual-arm. They utilized sliding mode control with a sliding perturbation observer. In another study by Tayfun et al. [25], the authors presented bilateral control of a tele-operated system consisting of one master and two-slave systems. The master system was a 6DOF haptic robot while one of the slave systems was a virtual 6DOF robot and another real industrial robot. A visual user interface was created to show the position and velocity profile for tele-operated control. They implemented Lyapunov stability methodology to analyze stability and position localization. Shafiqul et al. [26] investigated a bilateral control tele-operated scheme for a robotic system with unsymmetrical time-varying delays. The authors implemented an adaptive algorithm to determine relating factors between human and master manipulator and between slave and remote setup. Later, delayed estimated factors are fed back to the master and slave systems. The authors estimated and analyzed the impedance properties of the interaction between the human-based system and the remote environment. Further application of bilateral control are electro-pneumatic actuators and mobile robot [27,28].

Xiao et al. [29] presented the model mediated tele-operated approach. This proposed scheme has been designed to achieve stability and transparency while considering the random communication delays. Da Sun et al. [30], proposed a novel approach for tele-operated systems utilizing an extended prescribed performance control and a wave-based time domain passivity scheme. This scheme ensures synchronization of velocity, force, and localization. The stability and performance of the system are analyzed by the standard Lyapunov scheme. This methodology also ensures high tracking results of localization, velocity, and force. Similarly Azimifar et al. [31] presented a strategy to estimate the external force acting on master and slave systems. This scheme is advantageous with low-cost features and ease of implementation as the force sensors requirement is eliminated. They proposed a novel control for a nonlinear bilateral tele-operated system with time delays that estimates the force accordingly. The stability of the system is analyzed through the famous Lyapunov stability methodology. The tele-operation of remote system falls under the umbrella of bilateral control. It includes a master robot system that is operated by human operators and controlled through electric actuators whereas a slave robot system that is placed or installed at a remote location and is controlled through hydraulic actuators. Thus, to achieve an efficient bilateral control, three key features are necessary. The first is the coordinated control that is responsible for decoupling of position and force controllers. Second, is the linearization of hydraulic actuators. Third, is the agreement of system order of electric and hydraulic actuators by using a pseudo differentiator. Sho et al. [32] proposed a novel method with an efficient and accurate tracking performance and is the most stable in a contact control scenario among all three controllers.

Minou et al. [33] proposed a hybrid control algorithm for trajectory tracking with constant force. A non-linear model is implemented for the position controller to achieve predictive tracking, satisfying the input constraints. The authors analyzed the performance of the proposed controller and concluded that position tracking has a proof-mean-square-error (RMSE) of 0.89 mm whereas the catheter regulated the force with RMSE of 4.9 mN. Yaoyao et al. [34] presented a novel idea of trajectory control of an underwater vehicle-manipulator system by implementing a discrete time delay estimating methodology. Xia et al. [35] proposed a non-linear adaptive control algorithm for a non-linear manipulator system with uncertainties in its dynamics. Their algorithm guarantees accurate and efficient tracking. The proposed algorithm not only estimates the manipulator's dynamics but also determines the best fit dead-zone parameter in adaptation law. The estimated values are later used in proposed control law. The authors claimed and showed results suggesting that tracking is accurate and efficient even when both dynamics and dead-zone uncertainties appear at the same time. There are several studies in the literature which have presented the idea of utilizing force sensors to detect the outer forces in conventional bilateral control schemes, but such addition of force sensors has certain problems. A simple explanation to it is that a manipulator system with a force sensor acts as a two-mass resonant system. Thus, it is a bottleneck in realizing high-frequency force sensing [36,37]. The master and slave configuration of a manipulator system is highly applicable and suitable for nuclear power plants, because of restricted human access. In addition to this, it also requires exact control and dismantling/handling of a material/object with excessive load. Hydraulic systems are best suitable for such an application as they offer high power actuators.

In keeping a view of the above studies, in this study, we implemented the sliding mode control with sliding perturbation observer (SMCSPO). It is an efficient and robust control algorithm that not only estimates the reaction force of master and slave but also determines the bilateral control of the hydraulic servo system of a 3DOF master–slave robot. The reason for using a hydraulic servo system is that its power-to-weight ratio is better than any other type of actuated robot at the expense of positional accuracy. In this study, the sliding perturbation observer (SPO) is implemented to estimate the reaction force of the slave without using any sensor. The bilateral control scheme is implemented for efficient and accurate position and force tracking b/w master–slave configuration with visual feedback. In the bilateral controller, the difference of reaction force of the slave manipulator and operating force applied to the master manipulator is designed to target the impedance model. The reaction force of the slave is resultant of effects in the remote environment while operator force is applied by the operator (Human) at the master manipulator. The experimental results of studies endorse that the slave efficiently follows the position trajectory of the master system.

The manuscript is organized as follows: Section 2 of this paper presents the mechanical structure and dynamics of a 3DOF hydraulic servo system that could be utilized in the dismantling of nuclear power plants. Section 3 describes the sliding mode control with sliding perturbation observer (SMCSPO), and the reaction force estimation method is also presented in the same section. Section 4 is reserved for the description of mathematical details regarding bilateral control of master–slave 3DOF hydraulic servo system, and Section 5 presents its experimental setup. Section 6 presents the results of the performance of bilateral control through different experiments, and finally, Section 7 entails the concluding remarks to this study.

2. Mechanical Design and Dynamics of Hydraulic Servo System

It is a well-known fact that hydraulic servo systems have a favorable position in applications involving the machine tool industry. Some popular examples include handling hazardous material, remote equipment, steel factories, mining of materials, the exploration of oil, and the testing of automotive equipment, etc. A servo system is responsible, according to design, to control dynamical properties, such as; force, pressure, acceleration, electrical properties, etc. The attractive features of hydraulic servo systems include efficient response time, high torque and very few stroke characteristics. The hydraulic system has several advantages, but most prominent are accurate tracking of localization and acceleration, better stiffness features, null backlash, efficient response to sudden change, and less wear rate, among others. The hydraulic system has a better position among robotic systems to dismantle nuclear plants or hazardous plants. The hydraulic system, while used in nuclear plants, could be divided into two major parts, the first consists of a couple of hydraulic cylinders which formulate the vertical movement and the second consists of an AC servo motor responsible for horizontal changes. Two cylinders with a hydraulic mechanism is provided in the first part as much higher torque is required in the vertical direction. The mechanical design of shape and different sizes of a hydraulic servo system utilized in nuclear power plants is shown in Figures 1 and 2. The modelling of hydraulic actuator has been presented in Appendix A.

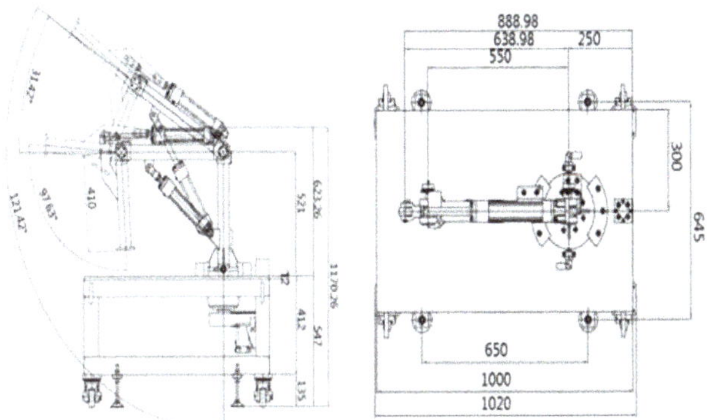

Figure 1. Structure of the 3DOF hydraulic servo system.

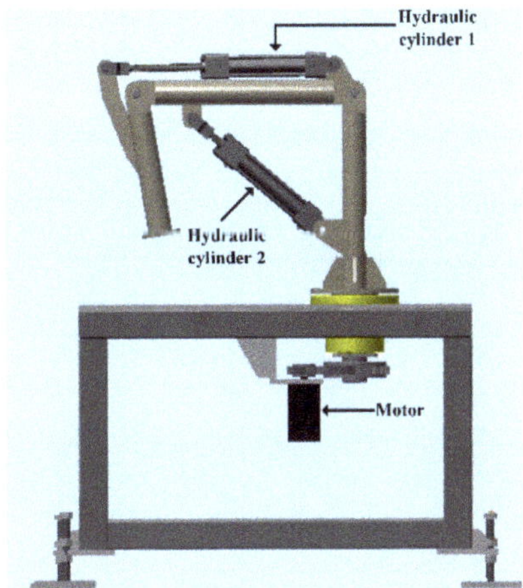

Figure 2. The 3DOF hydraulic servo system.

The schematic design of an end effector is displayed in Figure 3 and the schematic diagram of a hydraulic servo system's base and secondary link is shown in Figure 4.

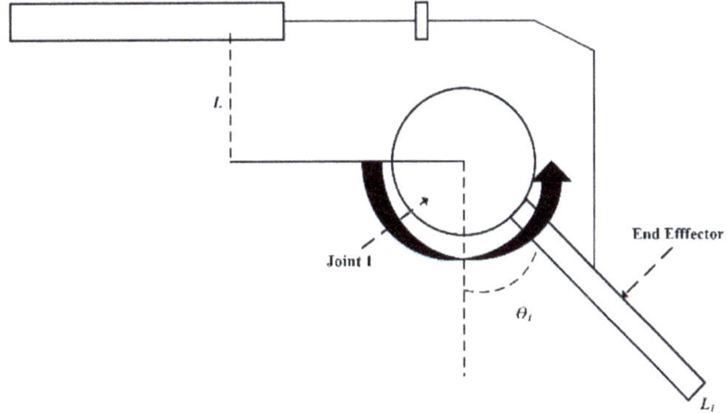

Figure 3. Schematic diagram of end effector.

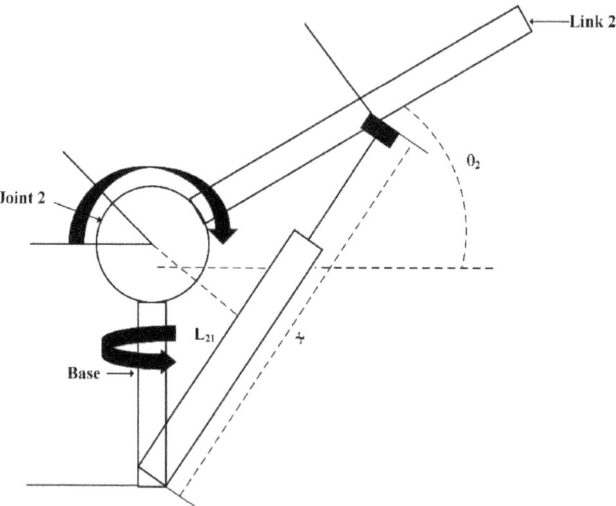

Figure 4. Schematic diagram of base and second link.

It is a well-known fact that a robotic system dynamic is a cross-relation between different forces, acceleration properties, and localization. Thus, the dynamical equation of a robotic manipulator in free space could be represented mathematically as,

$$T = A(\theta)\ddot{\theta} + B\left(\theta, \dot{\theta}\right) + g(\theta) \tag{1}$$

where θ, $A(\theta)$, $B\left(\theta, \dot{\theta}\right)$, $g(\theta)$ and T are the vectors representing joints of angles, the matrix representing mass or inertia, the centrifugal/Coriolis torque, the gravity torque in joint space, and the vector of joint torques, respectively. Similarly, the equation repressing dynamical properties of different links in a hydraulic servo system could be represented as

$$(J_{s1} + \Delta J_{s1})\ddot{\theta}_1 + (D_{s1} + \Delta D_{s1} + \beta_1)\dot{\theta}_1 + 0.5M_{s1}L_1 g \sin\theta_1 + \tau_{e1} = T_1 \tag{2}$$

$$(J_{s2} + \Delta J_{s2})\ddot{\theta}_2 + (D_{s2} + \Delta D_{s2})\dot{\theta}_2 + M_{s2}L_2 g \cos\theta_2 + \beta_2\dot{x} + \tau_{e2} + \lambda = T_2 \tag{3}$$

$$(J_{s3} + \Delta J_{s3})\ddot{\theta}_3 + (D_{s3} + \Delta D_{s3})\dot{\theta}_3 + 0.5M_{s3}L_3 g \sin\theta_3 + \tau_{e3} = T_3 \tag{4}$$

where J_{s1}, J_{s2} and J_{s3} represent the inertia of the base, second link, and end effector respectively, similarly D_{s1}, D_{s2} and D_{s3} represents the damping characteristics of the base, second link, and end effector respectively. The uncertainty is represented by Δ, β_1, and β_2 shows the viscosity of each cylinder in the first part. M_{s1}, M_{s2} and M_{s3} represent the masses of the base, second link and end effector, L_1 and L_3 are lengths of base and end effector, L_2 represents the length from joint to the centre of mass (COM) of the second link. τ_{e1}, τ_{e2} and τ_{e3} represent the reaction torque generated by contact with the environment and joints 1, 2 and 3 respectively, λ represents the dynamical properties regarding the base, $\dot{\theta}_1$ is the velocity of the first cylinder and $\dot{x}$ is the velocity of the second cylinder, and T_1, T_2 and T_3 represent joint torques of the base, second link and end effector respectively.

3. Sliding Mode Control with Sliding Perturbation Observer (SMCSPO)

3.1. Sliding Mode Control

Several previous studies have shown that the implementation of a sliding perturbation observer (SPO) and a sliding mode control (SMC) at the same time provides certain attractive features which include efficient performance against perturbation by utilizing partial state feedback. The combination of these two with such attractive properties is known as sliding mode control with sliding perturbation observer [37]. In our previous work [38], we only estimated the reaction force of the slave at the end effector. In this study, three-link robotic manipulator actuators are controlled by implementing SMC, and the reaction force is determined by applying SPO. A generic mathematical representation of n-degrees of freedom system inheriting second order dynamics is as follow,

$$\ddot{x}_j = f_j(\mathbf{x}) + \Delta f_j(\mathbf{x}) + \sum_{i=1}^{n} [(b_{ji}(x) + \Delta b_{ji}(\mathbf{x}))u_i] + d_j(t),\ j = 1, \ldots, n \tag{5}$$

where $\mathbf{x} \triangleq [\mathbf{X}_1 \ldots \mathbf{X}_\mathbf{n}]^T$ represents the state vector with $\mathbf{X_j} \triangleq [x_j, \dot{x}_j]^T$, the non-linear driving force is represented by $f_j(\mathbf{x})$ with uncertainties $\Delta f_j(\mathbf{x})$, the elements of the control gain matrix is represented by b_{ji} with corresponding uncertainties Δb_{ji}, the external disturbance and control input is represented by d_j and u_j; respectively, and f_j, b_{ji} are well known continuous functions of state in literature [39]. All of the uncertainties could be summed up, to represent perturbation as follow,

$$\psi_j(\mathbf{x}, t) = \Delta f_j(\mathbf{x}) + \sum_{i=1}^{n} [\Delta b_{ji}(\mathbf{x})u_i] + d_j(t) \tag{6}$$

The objective of control is to force state x to a desired state $\mathbf{x}_d \triangleq [\mathbf{X}_{1d} \ldots \mathbf{X}_{nd}]^T$ in the presence of perturbation. The upper bound for perturbation is defined by a known continuous function of state as follows

$$\Gamma_j(x, t) = F_j(x) + \sum_{i=1}^{n} |\Phi_{ji}(x)u_i| + D_j(t) > |\Psi_j(t)| \tag{7}$$

where $F_j > |\Delta f_j|$, $\Phi_{ji} > |\Delta b_{ji}|$ and $D_j > |d_j|$ represent the expected upper bounds of the uncertainties. Let us suppose $f_j(\mathbf{x})$, defined in Equation (5), except perturbation of Equation (6) is represented as

$$f_j(\hat{x}) + \sum_{i=1}^{n} b_{ji}(\hat{x})u_i = \alpha_{3j}\overline{u}_j \tag{8}$$

where α_{3j} is an arbitrary positive number, and $\overline{u}_j$ is the new control variable. The equations of SPO are derived as [39].

$$\dot{\hat{x}}_{1j} = \hat{x}_{2j} - k_{1j}sat(\tilde{x}_{1j}) \tag{9}$$

$$\dot{\hat{x}}_{2j} = \alpha_3\overline{u}_j - k_{2j}sat(\tilde{x}_{1j}) + \hat{\Psi}_j \tag{10}$$

$$\dot{\hat{x}}_{3j} = \alpha_{3j}^2(\overline{u}_j + \alpha_{3j}\hat{x}_{2j} - \hat{x}_{3j}) \tag{11}$$

$$\hat{\psi}_j = \alpha_{3j}(\alpha_3\hat{x}_{2j} - \hat{x}_{3j}) \tag{12}$$

where k_{1j}, k_{2j}, α_{3j} are positive numbers and $\tilde{x} = \hat{x} - x$ is the estimation error of the measurable state. $\hat{\psi}_j$ is defined as the estimated perturbation of the robot manipulator. The "~" and "^" represent the error in estimation and quantity estimated in result of the estimation respectively.

$$sat(\tilde{x}_{1j}) = \begin{cases} \tilde{x}_{1j}/|\tilde{x}_{1j}|, & if|\tilde{x}_{1j}| \geq \varepsilon_{0j} \\ \tilde{x}_{1j}/\varepsilon_{0j}, & if|\tilde{x}_{1j}| \leq \varepsilon_{0j} \end{cases} \tag{13}$$

The anti-chatter properties are formulated by a saturation function defined by Slotine et al. [40] where the boundary layer of SMC control is represented by ε_{0j}. Finally, the error dynamics of SPO are mathematically defined as [39],

$$\dot{\tilde{x}}_{1j} = \tilde{x}_{2j} - k_{1j}sat(\tilde{x}_{1j}) \tag{14}$$

$$\dot{\tilde{x}}_{2j} = -k_{2j}sat(\tilde{x}_{1j}) + \tilde{\Psi}_j \tag{15}$$

$$\dot{\tilde{x}}_{3j} = -\alpha_{3j}^2(\alpha_{3j}\tilde{x}_{2j} - \tilde{x}_{3j}) + \dot{\Psi}/\alpha_{3j} \tag{16}$$

After the observer sliding mode begins, $\tilde{x}_{2j}$ dynamics become

$$\dot{\tilde{x}}_{2j} + (k_{2j}/k_{1j})\tilde{x}_{2j} = \tilde{\Psi}_j \tag{17}$$

The frequency domain relation between $\tilde{\Psi}_j$ and Ψ_j is derived as

$$\tilde{\Psi}_j(p) = \frac{p[p^2 + (k_{1j}/\varepsilon_{0j})p + k_{2j}/\varepsilon_{0j}]}{p^3 + (k_{1j}/\varepsilon_{0j})p^2 + (k_{1j}/\varepsilon_{0j})p + \alpha_{3j}^2(k_{2j}/\varepsilon_{0j})}(-\Psi_j(p)) \tag{18}$$

and this is equivalent to a high-pass filter. The sliding function is defined as

$$s_j = \dot{e}_j + c_{1j}e_j \tag{19}$$

where $e_j = x_{1j} - x_{1dj}$ is the actual position tracking error, $c_{1j} > 0$. As the sliding surface is reached, we define $\dot{s}_j = 0$ and sliding control is defined as

$$\dot{s}_j = -K_j sat(s_j) \tag{20}$$

where robust gain is represented by K_j and is supposed to be positive. The control input u_j is mathematically defined as follows,

$$u_j = B^{-1}Col(\ddot{x}_{1dj} + f_j(\mathbf{x}) + c\dot{e}_j)_j - K_j sat(s_j) \tag{21}$$

Similarly, the estimated sliding function is represented mathematically as

$$\hat{s}_j = \dot{\hat{e}}_j + c_{j1}\hat{e}_j \tag{22}$$

where, the estimation error of localizing tracking is $\hat{e}_j = \hat{x}_{1j} - x_{1dj}$ and jth degree of freedom motion $[x_{1dj}\dot{x}_{1dj}]^T$ and $c_{j1} > 0$. Similarly, the estimation error in sliding function is mathematically represented by $\tilde{s}_j = \hat{s}_j - s_j$. The estimation error in sliding function can be calculated by using Equations (19) and (22) as

$$\tilde{s}_j = \dot{\tilde{x}}_{1j} + c_{j1}\tilde{x}_{1j} \tag{23}$$

The new control input $\overline{u}_j$ is designed such that it forces $\dot{\hat{s}}\hat{s} < 0$ outside of the prescribed manifold. The desired $\hat{s}_j$-dynamics is defined as

$$\dot{\hat{s}}_j = -K_j sat(\hat{s}_j) \tag{24}$$

Thus $\dot{\hat{s}}_j$ can be calculated as follows

$$\begin{aligned} \dot{\hat{s}}_j = \alpha_{3j}\overline{u}_j - \left[k_{2j}/\varepsilon_{0j} + c_{j1}(k_{1j}/\varepsilon_{0j}) - (k_{1j}/\varepsilon_{0j})^2\right]\tilde{x}_{1j} \\ -(k_{1j}/\varepsilon_{0j})\tilde{x}_{2j} - \ddot{x}_{1jd} + c_{j1}\left(\hat{x}_{2j} - \dot{x}_{1jd}\right) + \hat{\psi}_j \end{aligned} \tag{25}$$

The control law to apply Equation (23) with $\widetilde{x}_{2j} = 0$ is defined as

$$\begin{aligned} \overline{u}_j = \tfrac{1}{\alpha_{3j}}\{-K_j sat(\hat{s}_j) + [\tfrac{k_{2j}}{\varepsilon_{0j}} + c_{j1}\tfrac{k_{1j}}{\varepsilon_{0j}} - (\tfrac{k_{1j}}{\varepsilon_{0j}})^2]\widetilde{x}_{1j} \\ +\ddot{x}_{1jd} - c_{j1}(\hat{x}_{2j} - \dot{x}_{1jd}) - \hat{\psi}_j\} \end{aligned} \tag{26}$$

where $\overline{u}_j$ represents the control input of SMCSPO. Thus, the resulting $\hat{s}_j$-dynamics including the effects of $\widetilde{x}_{2j}$ could be mathematically represented as

$$\dot{\hat{s}}_j = -K_j sat(\hat{s}_j) - (k_{1j}/\varepsilon_{0j})\widetilde{x}_{2j} \tag{27}$$

To ensure the outer part of the manifold $|\hat{s}_j| \leq \varepsilon_{0j}$ satisfies the inequality $\dot{\hat{s}}\hat{s} < 0$, the robust control gain must be constrained to the following inequality

$$K_j \geq k_{1j}^2/\varepsilon_{0j} \tag{28}$$

Finally, the actual s_j-dynamics within the boundary layer of $|\hat{s}_j| \leq \varepsilon_{0j}$ can be mathematically expressed as

$$\begin{aligned} \dot{s}_j + \tfrac{K_j}{\varepsilon_{0j}} s_j = \left[\tfrac{k_{2j}}{\varepsilon_{0j}} - \left(\tfrac{k_{1j}}{\varepsilon_{0j}} - \tfrac{K_j}{\varepsilon_{0j}}\right)\left(c_{j1} - \tfrac{k_{1j}}{\varepsilon_{0j}}\right)\right]\widetilde{x}_{1j} \\ -\left(c_{j1} + \tfrac{K_j}{\varepsilon_{0j}}\right)\widetilde{x}_{2j} - \widetilde{\psi}_j \end{aligned} \tag{29}$$

The estimation errors in state estimation and perturbation are the driving force for s_j-dynamics. The target of the designed SMCSPO is to minimize mismatch b/w the actual and desired trajectory. The sliding perturbation observer is responsible for reducing the estimation error $\widetilde{x}_j$ in the observer part. Therefore, the sliding mode control is designed for error $\hat{e}_j$ reduction b/w desired and actual values of the trajectory. Thus, implementation of SMC and SPO simultaneously guarantee the accurate results of trajectory tracking with a reduction of error.

The mechanical hardware limitations restrict the design procedure as described in Jairo et al. [39]. The instant when $|\hat{s}_j| \leq \varepsilon_{0j}$, the observer design and s_j-dynamics, can be represented in mathematical form as below,

$$\begin{bmatrix} \dot{\widetilde{x}}_{1j} \\ \dot{\widetilde{x}}_{2j} \\ \dot{\widetilde{x}}_{3j} \\ \dot{s}_j \end{bmatrix} = \begin{bmatrix} -k_{1j}/\varepsilon_{0j} & 1 & 0 & 0 \\ -\frac{k_{2j}}{\varepsilon_{0j}} & \alpha^2{}_{3j} & -\alpha_{3j} & 0 \\ 0 & \alpha^3{}_{3j} & -\alpha^2{}_{3j} & 0 \\ k_{2j}/\varepsilon_{0j} - (c - k_{1j}/\varepsilon_{0j})^2 & -(2c + \alpha^2{}_{3j}) & \alpha_{3j} & -c \end{bmatrix} \cdot \begin{bmatrix} \widetilde{x}_{1j} \\ \widetilde{x}_{2j} \\ \widetilde{x}_{3j} \\ s_j \end{bmatrix} + \begin{bmatrix} 0 \\ 0 \\ 1 \\ 0 \end{bmatrix} \dot{\psi}_j/\alpha_{3j} \tag{30}$$

where a square matrix of order 4 in Equation (30) represents the state matrix. Further suppose, λ represents the eigen values of state matrix A, then its characteristic equation det $|\lambda I - A| = 0$, can be expressed as,

$$[\lambda + c_{j1}]\left[\lambda^3 + (k_{1j}/\varepsilon_{0j})\lambda^2 + (k_{2j}/\varepsilon_{0j})\lambda + \alpha^2{}_{3j}(k_{2j}/\varepsilon_{0j})\right] = 0 \tag{31}$$

By implementing the pole-placement method, let us introduce a desired characteristic polynomial $p(\lambda_d) = (\lambda + \lambda_d)^4$ which leads to a design solution

$$k_{1j}/\varepsilon_{0j} = 3\lambda_d k_{2j}/k_{1j} = \lambda_d \alpha_{3j} = \sqrt{\lambda_d/3c} = K_j/\varepsilon_{0j} = \lambda_d \tag{32}$$

It is obvious that a reduction in error and an increase in the accuracy of observations could be achieved with a large value of gain λ_d. But Jairo et al. [41] has categorically shown the limitation of breakpoints of sliding function dynamics encircled in a manifold. They showed that it could not go beyond $1/5\tau^{hw}$, where h is a positive number, and frequency is represented by w. Another study by Slotine et al. [42] also showed the same results. Thus, the optimal gain value could be selected as

$$\lambda_d = \frac{1}{15\tau^{hw}} \tag{33}$$

3.2. Reaction Force Estimation Based upon Sliding Perturbation Observer (SPO)

The algorithm using SPO could be utilized to set an estimate of perturbation that leads to the determination of reaction force. The perturbation's estimation includes two factors, the first disturbance that is regarded as an external force and another is the dynamic error that inherits non-linear terms and viscous friction. The 3DOF robotic-manipulator defined in Equations (2) and (3) can be used to define a perturbation estimate that would be used to determine the reaction force. The perturbation estimate of the end-effector and second link are mathematically represented by an expression defined below,

$$\hat{\psi}_{s1} = -\frac{1}{J_{s1}}(\hat{\tau}_{e1}) - \frac{1}{J_{s1}}(0.5M_{s1}L_1 g \sin\theta_1) - (\frac{\Delta J_{s1}}{J_{s1}})\ddot{\theta}_1 - \frac{1}{J_{s1}}(\Delta B_{s1} + \beta_1)\dot{\theta}_1 \tag{34}$$

$$\hat{\psi}_{s2} = -\frac{1}{J_{s2}}(\hat{\tau}_{e2}) - \frac{1}{J_{s2}}(M_{s2}L_2 g \cos\theta_2) - (\frac{\Delta J_{s2}}{J_{s2}})\ddot{\theta}_1 - \frac{1}{J_{s2}}(\Delta B_{s2}\dot{\theta}_2) - \frac{1}{J_{s2}}(\beta_2\dot{x}) - \frac{1}{J_{s2}}(\lambda) \tag{35}$$

The Equations (34) and (35) further used to calculate the reaction force as follows

$$\hat{\tau}_{e1} = J_{s1}\hat{\psi}_{s1} + 0.5M_{s1}L_1 g \sin\theta_1 + \Delta J_{s1}\ddot{\theta}_1 + (\Delta B_{s1} + \beta_1)\dot{\theta}_1 \tag{36}$$

$$\hat{\tau}_{e2} = J_{s2}\hat{\psi}_{s2} + M_{s2}L_2 g \cos\theta_2 + \Delta J_{s2}\ddot{\theta}_1 + \Delta B_{s2}\dot{\theta}_2 + \beta_2\dot{x} + \lambda \tag{37}$$

where $\hat{\tau}_{e1}$, $\hat{\tau}_{e2}$ are the estimates of reaction torques generated as a result of contact with the environment of the end-effector and second link respectively, Δ represents the uncertainty parameter. It is worthy to mention that if the parameter is well estimated than the parameter of uncertainty could be considered as well.

4. Bilateral Control Design between Master and Slave for 3DOF Hydraulic Servo System

4.1. Bilateral Control

The design purpose of bilateral control refers to a relation b/w the master and slave system helping the system operator in a way that is as close as the actual environment. There are two important factors that need to be considered while designing bilateral control. The first is that the slave accurately follows the trajectory of the master system and the second that the system's operator gets a realistic feel of a reaction force. In this study, the master and slave system are 3DOF hydraulic servo systems. The dynamical equations for these two systems are shown in Equations (38) and (39).

$$J_m\ddot{\theta}_m + B_m\dot{\theta}_m = u_m + \tau_h \tag{38}$$

$$J_s\ddot{\theta}_s + B_s\dot{\theta}_s = u_s - \tau_e \tag{39}$$

where J_m, J_s, u_m and u_s are inertia, position, and control input of master and slave of hydraulic servo system respectively. τ_h represents the real/action force generated by the operator at the master device, τ_e defines the reaction force of the slave system in a remote environment. A detailed workflow of the control algorithm for bilateral control is presented in Figure 5.

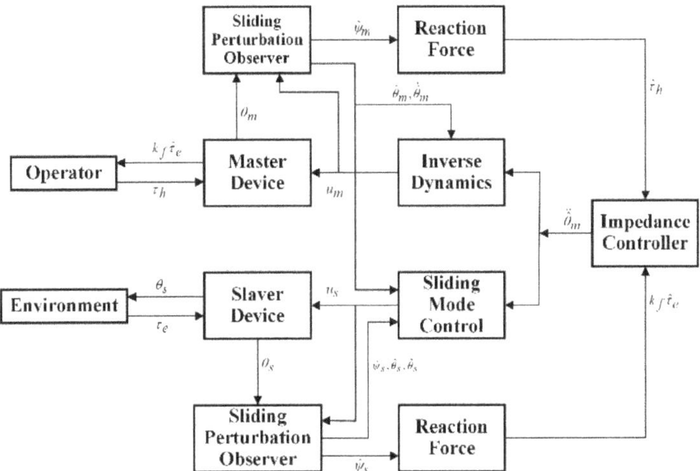

Figure 5. The work flow of control algorithm.

A simple fact is that as the master device is operated by a human operator, the slave system follows the trajectory of the master system. The slave system of hydraulic servos follows the master's trajectory using a sliding mode control. The impedance control is implemented to transfer reaction force during trajectory tracking. Therefore, no reaction force is observed by the operator while there are no contact between the slave and environment.

4.2. Master Controller and Device

The design of the master system has been considered under two constraints. The first one is that the reaction force is felt by the operator during the slave system and environment contact. The second constraint is that the operator uses the master system with the minimal force possible. The design of the impedance control is defined as,

$$J_d\ddot{\theta}_m + B_d\dot{\theta}_m + K_d\theta_m = \hat{\tau}_h - k_f\hat{\tau}_e \tag{40}$$

where J_d, B_d and K_d are the inertia, damping, and stiffness of the impedance control model respectively, k_f represents the scale factor of the reaction force. The scale factor k_f scales the force ratio that is converted from the master system to the slave system. For simplicity, in this study k_f is considered to be unity. The control input signal to the master system can be determined by replacing the estimated/observed state variable to the original ones. Thus, the control input for a dynamical system of Equation (38) and impedance control of Equation (40) can be mathematically presented as follows,

$$u_m = (B_m - \frac{J_m}{J_d}B_d)\hat{\dot{\theta}}_m + (\frac{J_m}{J_d} - 1)\hat{\tau}_h - \frac{J_m}{J_d}(k_f\hat{\tau}_e + K\hat{\theta}_m) \tag{41}$$

where u_m, $\hat{\dot{\theta}}_m$, $\hat{\theta}_m$ are the control input, estimated speed and estimated position of the master system by utilizing SPO, $\hat{\tau}_h$ represents the torque estimate that is produced by the operator while operating the master device. The master device used in this research is presented in Figure 6.

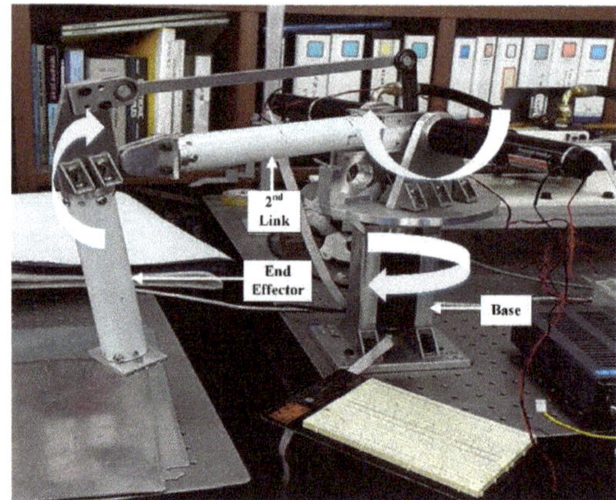

Figure 6. Master device.

4.3. Slave Controller and Device

The slave controller is designed to track the trajectory of the master system. The slave controller is designed by utilizing SMCSPO for this study. The logic behind the SMCSPO design is to compensate sliding mode control for the perturbation estimate. It is an additive quantity consisting of uncertainty, disturbance, and non-linearity [39]. Thus, SMCSPO design efficiently removes the external disturbance and uncertainty of parameters. The sliding function that is estimated can be expressed as

$$\hat{s} = \dot{\hat{e}} + c\hat{e} \tag{42}$$

where $\hat{e} = \hat{\theta}_{s1} - \hat{\theta}_{m1}$ is the tracking error between the master and slave device, c is a positive constant. By substituting the observed/estimated state in the sliding function defined above in Equation (42) and the sliding perturbation observer of Equation (12), the differential of the estimated sliding surface $\hat{s}$ can be mathematically represented as

$$\begin{aligned}\dot{\hat{s}} = \alpha_{s3}\overline{u}_s - \tfrac{k_{s2}}{\varepsilon_{s0}}\widetilde{\theta}_{s1} - \alpha_{s2}\widetilde{\theta}_{s1} + \hat{\psi}_s - \tfrac{k_{s1}}{\varepsilon_{s0}}\dot{\widetilde{\theta}}_{s1} \\ -\alpha_{s1}\dot{\widetilde{\theta}}_{s1} - \ddot{\theta}_{m1} + c(\dot{\hat{\theta}}_{s1} - \dot{\hat{\theta}}_{m1})\end{aligned} \tag{43}$$

It is found that $\alpha_{s1}\dot{\widetilde{\theta}}_{s1}, \alpha_{s2}\widetilde{\theta}_{s1} \approx 0$ when the phase is converged to the sliding surface and error estimate $\widetilde{\theta}_{s1}$ remains inside boundary layer. The impedance model Equations (40) and (43) are utilized to obtain $\dot{\hat{s}}$.

$$\begin{aligned}\dot{\hat{s}} = \alpha_3\overline{u}_s - [\tfrac{k_{s2}}{\varepsilon_{s0}} + c(\tfrac{k_{s1}}{\varepsilon_{s0}}) - (\tfrac{k_{s1}}{\varepsilon_{s0}})^2]\widetilde{\theta}_{s1} - J_d^{-1}(\hat{\tau}_h \\ -k_f\hat{\tau}_e - B_d\dot{\hat{\theta}}_m - K_d\hat{\theta}_m) + \hat{\psi}_s + c(\dot{\hat{\theta}}_{s1} - \dot{\hat{\theta}}_{m1})\end{aligned} \tag{44}$$

The new control variable $\overline{u}_s$ is chosen under the constraint i.e., $\dot{\hat{s}}\hat{s} < 0$. Similarly, $\hat{s}$ dynamics is chosen to satisfy the sliding mode condition

$$\dot{\hat{s}} = -Ksat(\hat{s}) \tag{45}$$

The new control input presented in Equation (8) can be found from Equations (44) and (45)

$$\begin{aligned}\overline{u}_s = \alpha_{s3}^{-1}\{-Ksat(\hat{s}) + [\tfrac{k_{s2}}{\varepsilon_{s0}} + c(\tfrac{k_{s1}}{\varepsilon_{s0}}) - (\tfrac{k_{s1}}{\varepsilon_{s0}})^2]\widetilde{\theta}_{s1} \\ +J_d^{-1}(B_d\dot{\hat{\theta}}_m + K_d\hat{\theta}_m - \hat{\tau}_h + k_f\hat{\tau}_e) - c(\dot{\hat{\theta}}_{s1} - \dot{\hat{\theta}}_{m1}) - \hat{\psi}_s\}\end{aligned} \tag{46}$$

Final control input for the slave system is mathematically defined as

$$u_s = J_s\overline{u}_s + B_s\dot{\hat{\theta}}_s \tag{47}$$

5. Experimental Environment

The slave device for the experiment is shown in Figure 7. The hydraulic servo system consists of two hydraulic cylinders and one AC servo motor. The first base axis is actuated by the AC servo motor. The 2nd link and end effector are actuated by the hydraulic cylinder.

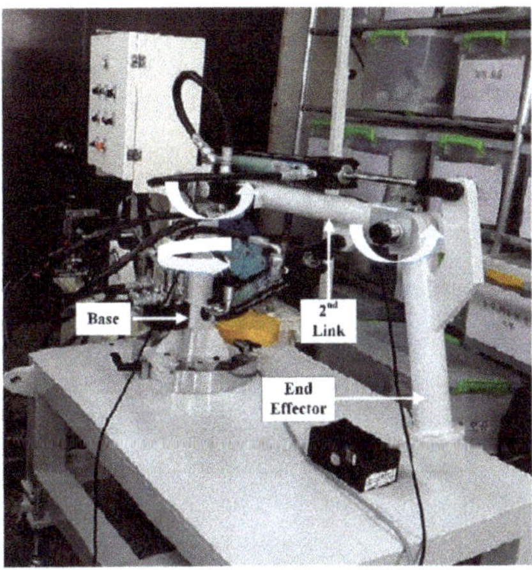

Figure 7. Slave device.

The experiments were performed on a 3DOF hydraulic servo system. Table 1, shows the specifications of a hydraulic servo system.

Table 1. Hydraulic servo system.

S. No	Items	Specification
1	Hydraulic cylinder	Piston and rod diameter = 0.04 m, 0.022 Stroke = 20 cm
2	Hydraulic pump	P_max = 210 bar Q_max = 20 1/min
3	Displacement transducer	Stroke = 20 cm (10 V)
4	Propositional directional control valve	D633-313A, Moog, Inc.
5	Relief valve	P_set = 20 bar
6	Control board	PC based MMC

In [43,44], the identification and robust control of a hydraulic servo system have been discussed. The unknown parameters such as the inertia and damping coefficients of the system are obtained by the signal compression method which can estimate the dynamics by obtaining an equivalent impulse response [45]. The model dynamic is given by

$$\ddot{x} = \frac{1}{J_i}u - \frac{1}{J_i}D_i\dot{x}, \quad i = 1,\ 2,\ 3 \tag{48}$$

where J_i, D_i are the equivalent moments of inertia and damper respectively. Table 2, shows the values of the dynamics.

Table 2. Hydraulic servo system dynamics parameters.

S. No	Moment of Inertia (Master) kg·m^2	Moment of Inertia (Slave) kg·m^2	Damper (Master) kg·m^2	Damper (Slave) kg·m^2
1	1.35135	303.26	3.99	17,355.5
2	1.5	59.52	3.99	5241.66
3	0.74	355.91	3.99	2214

The experimental setup includes a master device, a slave device and a control system as shown in Figure 8.

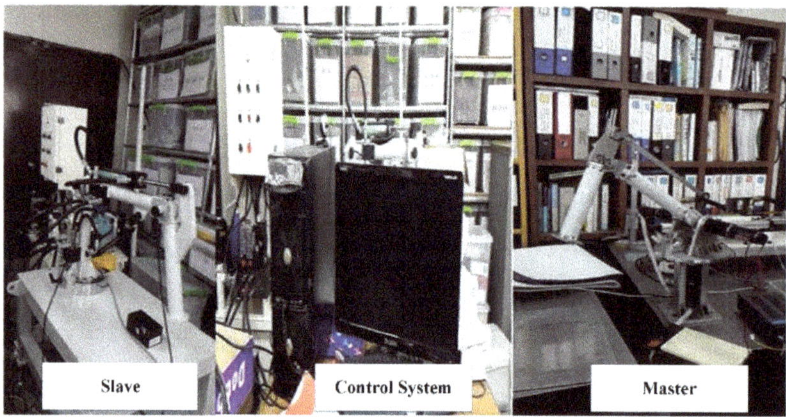

Figure 8. Experimental setup.

The master and slave device consist of three links each, in which the third link is connected with the base. We can find the reaction force at end effector and the 2nd link. SMCSPO is used to estimate the reaction force. Since the operator moves the master device to make the slave come in contact with a hard object, a visual sensor is installed with the slave device. It is not relevant to the control algorithm but provides visual feedback to the operator to avoid any inconvenience or uncertain situations. Figure 9 provides a pictorial view of visual feedback which helps enhance the safety during work.

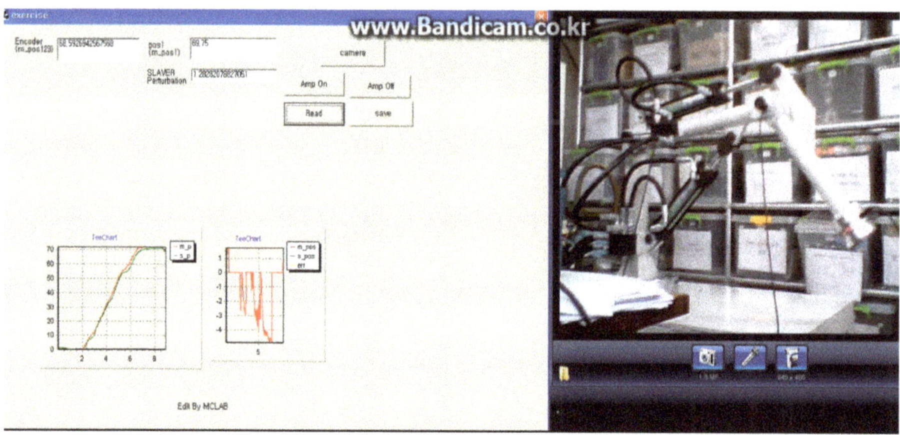

Figure 9. Pictorial view of visual feedback.

6. Experimental Results

In our study, our system is non-linear, and therefore, a robust control scheme (i.e., SMCSPO) was pursued. Since PID/PD controllers are not robust, comparatively to SMCSPO, to nonlinearities. Therefore, their performance is not as good as that of the proposed SMCSPO. Figure 10 shows the comparison between PID and SMCSPO.

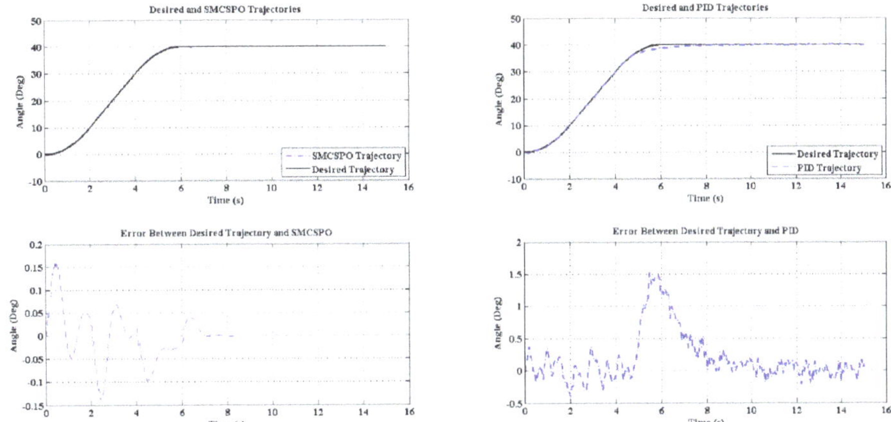

Figure 10. Comparison between PID and SMCSPO.

Table 3 shows the values of the parameters used in the experiment by using SMCSPO. The parameters used in the experiment are determined by using the Equations (32) and (33). Extensive simulations were performed to find the best value of boundary layer width (ε_o) resulting in convergence of the sliding surface to zero.

Table 3. Design parameters.

S. No	Parameters	Values
1	$k \cdot$ (End Effector)	25
2	$k \cdot$ (2nd Link)	250
3	$k \cdot$ (Base)	8
4	k_1	39
5	k_2	507
6	ε_0	1
7	c	13
8	e	1
9	α_3 (End Effector)	4.08
10	α_3 (2nd Link)	10
11	α_3 (Base)	2.58

Six different experiments are done by using SMCSPO: (i) Bilateral (Master–Slave manipulation) control of the end effector (ii), estimated perturbation measured at the end effector of master and slave (iii), bilateral (Master–Slave manipulation) control of the 2nd link (iv), estimated perturbation measured at the 2nd link of master and slave (v), bilateral (Master–Slave manipulation) control of the base (vi), and bilateral (Master–Slave manipulation) control of the end effector, 2nd link and base at the same time.

The scenario of the experiment is set up such that the operator (human) moves the master device and the slave device follows the trajectory of the master device using SMCSPO. Figure 11 shows the experimental result of the master–slave trajectories for the end effector. The blue line shows the experimental result of the master device whereas, the red line shows the experimental result of the slave device by using SMCSPO. It can be observed that the slave device can follow the master device. The maximum value of the trajectory is 81 degree at 14 s. The end effector of the slave device can move between 0 and 90 degrees.

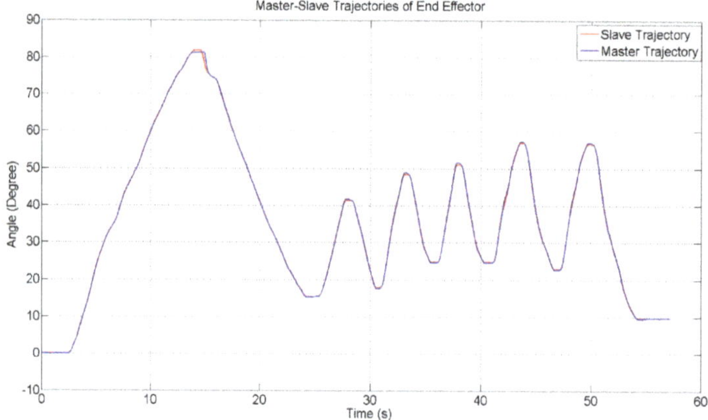

Figure 11. Master–slave trajectories for end effector.

The error between the master–slave trajectories for the end effector is shown in Figure 12. The maximum error between the master and slave trajectories is 0.62 degrees at 15.5 s. It can be seen that the errors between the master and slave trajectories are very small.

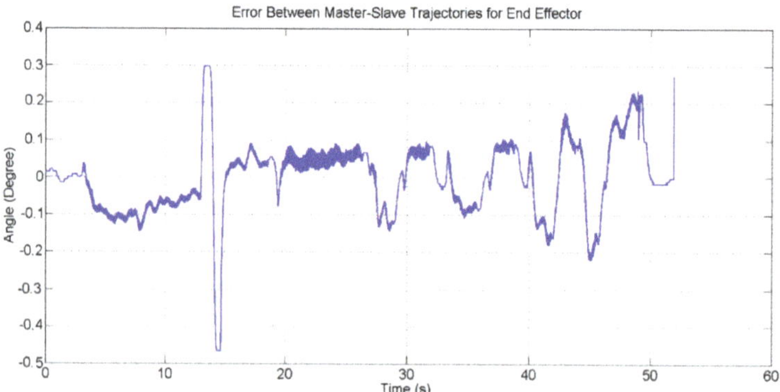

Figure 12. Error between master–slave trajectories for end effector.

Figures 13 and 14 show the estimated perturbation of master and slave for the end effector. The maximum estimated perturbation of the master is 67.56162 N*m at 14 s. While the maximum estimated perturbation of the slave is 1516.7 N*m at 14 s.

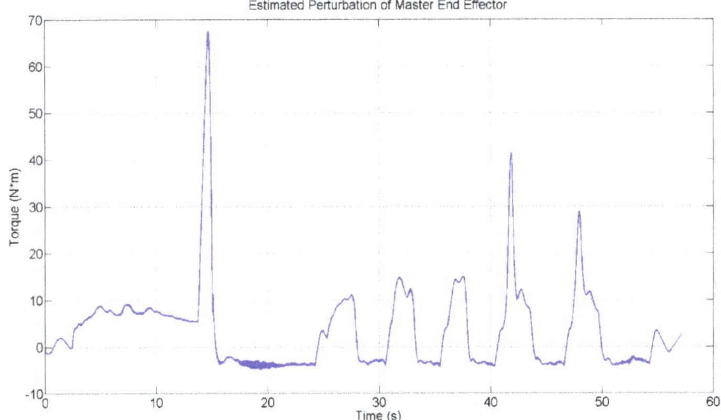

Figure 13. Estimated perturbation of master for end effector.

The value of the estimated perturbation for the slave is very high as compared to the master. This is because of the hydraulic system for the slave. The dynamic value of the slave (i.e., 303.26) is much bigger than the master (i.e., 1.35135). The pattern of the estimated perturbation of master and slave is the same but in an opposite direction. To compare these two results, we normalized the estimated perturbation of master and slave. The normalized estimated perturbation of master is between 0 and 1 while the slave is between −1 and 0 using Equations (49) and (50).

$$P_{norm}(Master) = \frac{a_i - \min(a)}{\max(a) - \min(a)}, \quad i = 1 \ldots N \tag{49}$$

$$P_{norm}(Slave) = \frac{a_i - \max(a)}{\max(a) - \min(a)}, \quad i = 1 \ldots N \tag{50}$$

In the above equations, a_i is the current value. Figure 15 shows the normalized estimated perturbation for the master and slave of the end effector. The red line shows the normalized estimated perturbation of the master device, whereas the blue line shows the normalized estimated perturbation of the slave device.

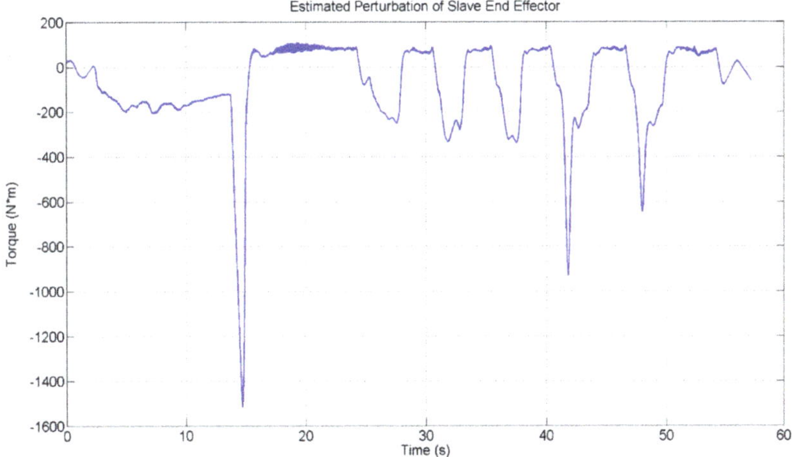

Figure 14. Estimated perturbation of slave for end effector.

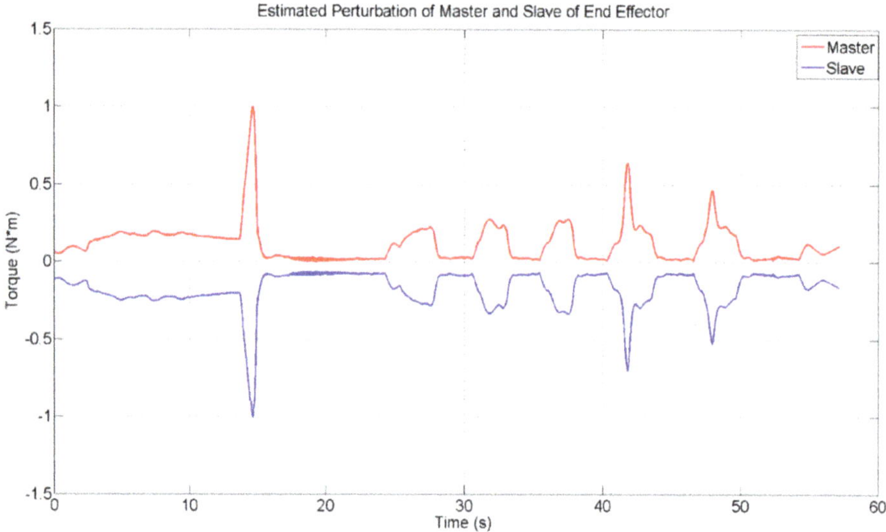

Figure 15. Normalized estimated perturbation of master and slave for end effector.

Figure 16 shows the experimental result of the master–slave trajectories for the 2nd link. The blue line shows the experimental result of the master device, whereas the red line shows the experimental result of the slave device by using SMCSPO. The slave device can also follow the master device. The maximum value of the trajectory is 60 degrees at 13.5 s. The 2nd link of the slave device can move between 0 and 90 degrees.

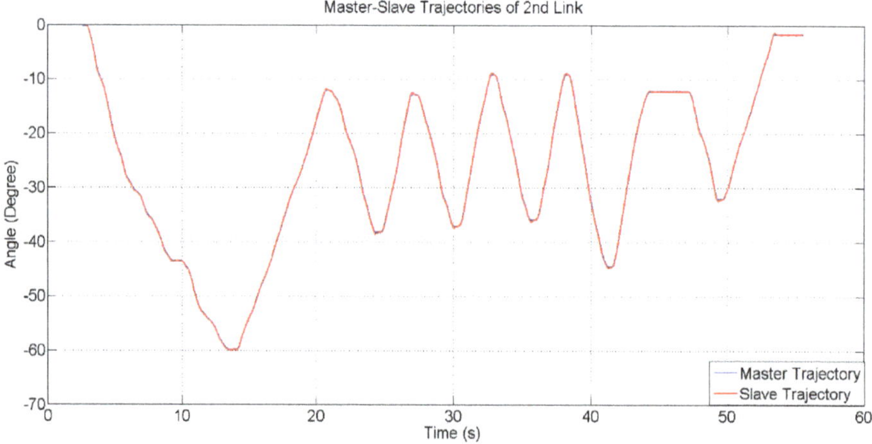

Figure 16. Master–slave trajectories for 2nd link.

The error between master–slave trajectories for the 2nd Link is shown in Figure 17. The maximum error between master and slave trajectories are 0.41 degree at 49 s.

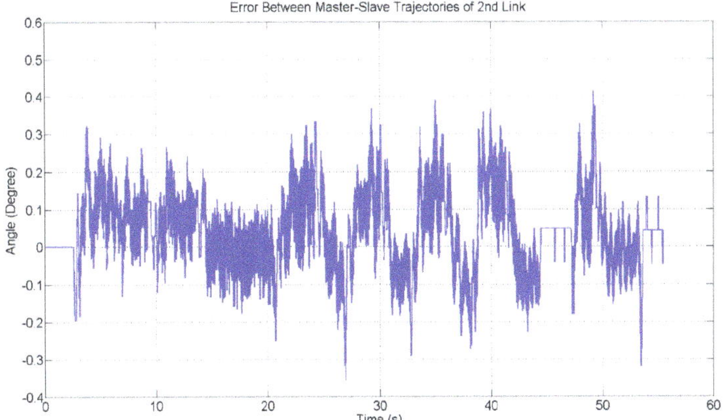

Figure 17. Error between master–slave trajectories for 2nd link.

Figures 18 and 19 show the estimated perturbation of master and slave for the 2nd link. The maximum estimated perturbation of the master is 326.6367 N*m at 38.5 s. While the maximum estimated perturbation of the slave is 1296.9 N*m at 38.5 s. The pattern of master and slave estimated perturbation is the same but in opposite direction.

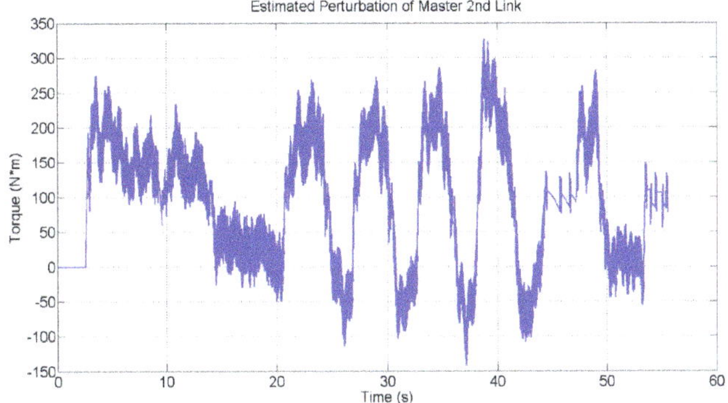

Figure 18. Estimated perturbation of master for 2nd link.

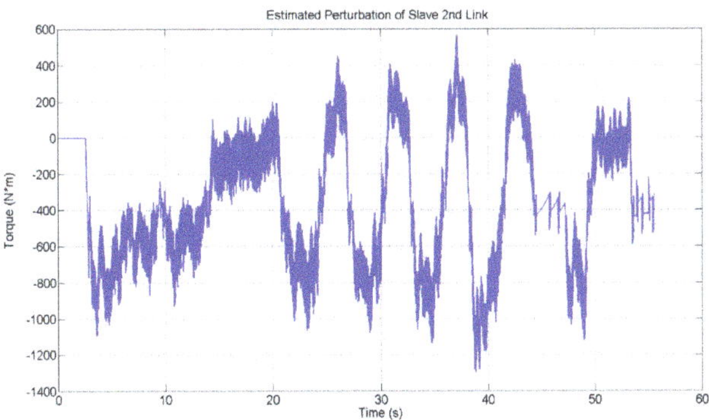

Figure 19. Estimated perturbation of slave for 2nd link.

Figure 20 shows the normalized estimated perturbation for master and slave of the 2nd link. The red line shows the normalized estimated perturbation of the master device, whereas the blue line shows the normalized estimated perturbation of the slave device.

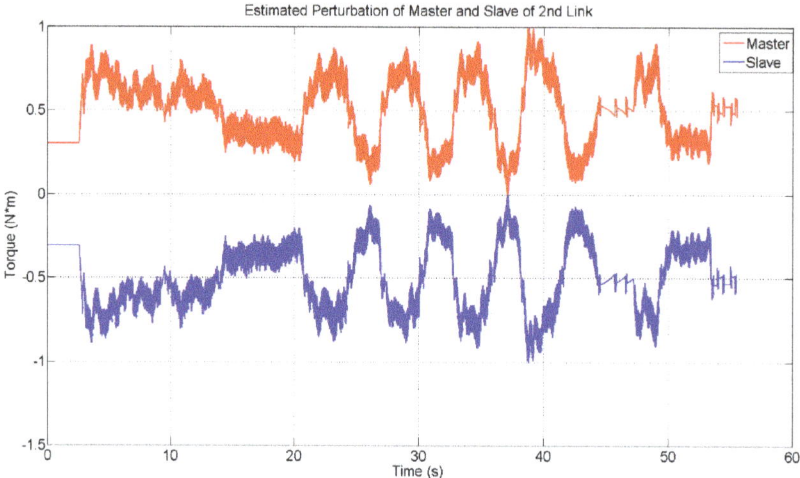

Figure 20. Normalized estimated perturbation of master and slave for 2nd link.

Figure 21 shows the experimental result of master–slave trajectories for the base axis. The blue line shows the experimental result of the master device whereas, the red line shows the experimental result of the slave device by using SMCSPO. It can be seen that the slave device can follow the master device. The maximum value of the trajectory is 100 degrees at 13 s. The base of the slave device can move between 0 and 360 degrees.

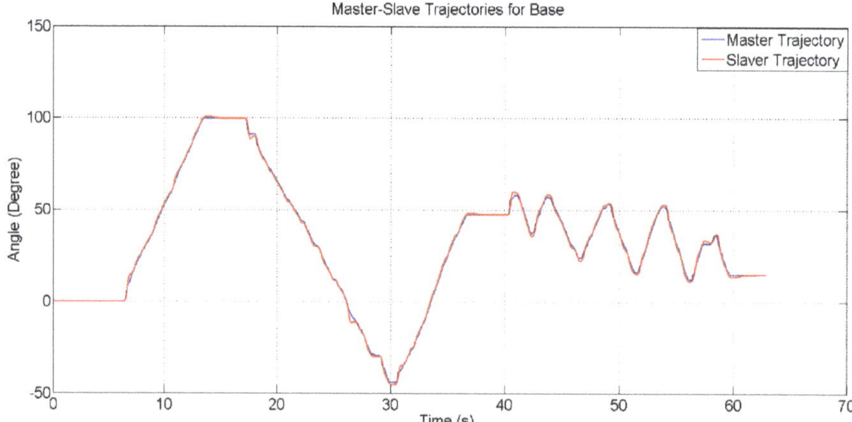

Figure 21. Master–slave trajectories for base.

The error between master–slave trajectories for the base is shown in Figure 22. The maximum error between master and slave trajectories is 1.3 degrees at 26.5 s.

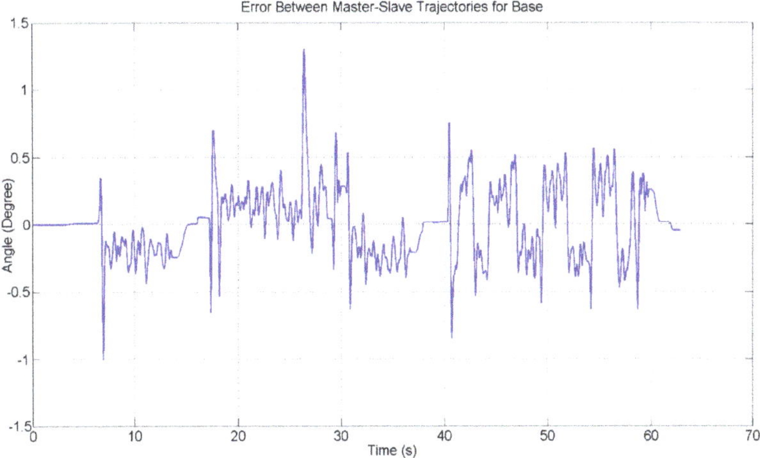

Figure 22. Error between master–slave trajectories for base.

Figure 23 shows the experimental result of the master–slave trajectories for the end effector, 2nd link and base by using SMCSPO. The blue line shows the experimental result of the master device whereas the red line shows the experimental result of the slave device of the end effector. The grey line shows the experimental result of the master device whereas the pink dotted line indicates the experimental result of the slave device of the 2nd link. The blue-dotted line shows the experimental result of the master device whereas the red dotted line shows the experimental result of the slave device of the base. At the beginning only, the end effector can move until 49 s. Then, the end effector and 2nd link moved simultaneously by 12 s (i.e., from 50 to 62 s). After that, the 2nd link moved separately by 18 s (i.e., from 62 to 80 s). Then, the base can move by 32 s (i.e., from 81 to 113 s). In the end, the end effector and 2nd link moved again simultaneously by 16 s (i.e., from 115 to 131 s). The slave device followed the trajectory of the master device accurately. The maximum trajectories of the end effector, 2nd link and base are 88, 27 and 52 degrees, respectively.

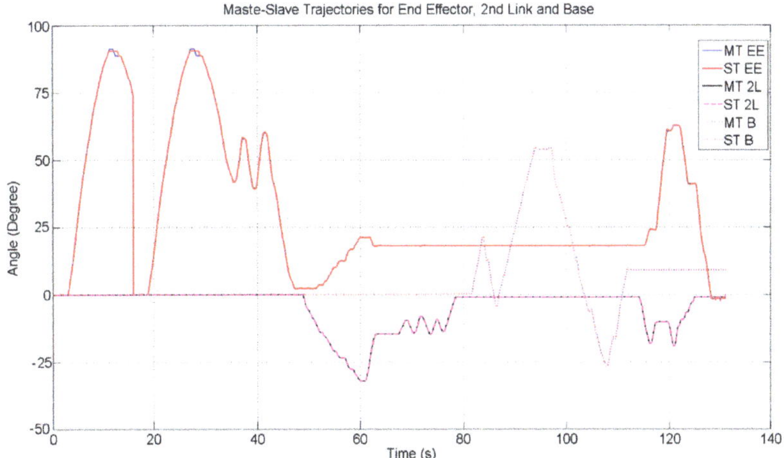

Figure 23. Master–slave trajectories for end effector, 2nd link and base.

7. Conclusions

In this paper, we have estimated the reaction force of the end effector and 2nd link for a three-degree of freedom hydraulic servo system with master–slave manipulators using SMCSPO without using any sensors. By using an SMC-based bilateral control strategy and visual feedback, the slave device followed the trajectory of the master device (human operator) with minimum error. Also, bilateral control is used to estimate the reaction force of the master device which is fed back to the operator to handle the master device.

From Figures 11, 16, 21 and 23, it is confirmed that the slave can follow the master trajectories and an operator can easily handle the slave device by feeling the estimated reaction force using the applied SPO when the slave touches an object even if a force sensor is not used.

The maximum error between master and slave for the end effector, 2nd link and base are summarized in Table 4.

Table 4. Error between master–slave trajectories.

S. No	Links	Maximum Error (Degree)	Maximum Trajectory (Degree)
1	End effector	0.62	81
2	Link 2	0.4	60
3	Base	1.3	100

This research is applied for dismantling nuclear power plants, and there are many situations where a human cannot access due to the high degree of radiation and the very long half-lives of the radioactive materials involved. Therefore, the slave device is used for such hazardous locations. It is useful for activities such as transportation of active uranium in nuclear power plants, disposal of an explosive, remote cutting for nuclear power plant dismantling, grinding, etc.

Author Contributions: Conceptualization, K.D.K. and M.C.L.; Formal analysis, K.D.K., J.W. and S.J.A.; Funding acquisition, M.C.L.; Methodology, K.D.K.; Software, K.D.K. and S.J.A.; Supervision, M.C.L.; Writing—original draft, K.D.K.; Writing—review & editing, K.D.K., J.W. and M.C.L.

Funding: This research was funded by the MOTIE (Ministry of Trade, Industry & Energy), Korea, under the Industry Convergence Liaison Robotics Creative Graduates Education Program supervised by the KIAT (N0001126). This research was funded by the Technology Innovation Program (10073147, Development of Robot Manipulation Technology by Using Artificial Intelligence) funded By the Ministry of Trade, Industry & Energy (MOTIE, Korea).

Acknowledgments: This research was funded by the MOTIE (Ministry of Trade, Industry & Energy), Korea, under the Industry Convergence Liaison Robotics Creative Graduates Education Program supervised by the KIAT (N0001126). This research was funded by the Technology Innovation Program(10073147, Development of Robot Manipulation Technology by Using Artificial Intelligence) funded By the Ministry of Trade, Industry & Energy(MOTIE, Korea).

Conflicts of Interest: The author declares no conflict of interest.

Appendix

Figure A1 is showing the schematic of hydraulic servo system. In this figure, V_e is the volume of the chamber of the cylinder, K_{sv} is coefficient between the input voltage and displacement of spool,x_v is the displacement of servo valve spool and M_p is the mass factor of the end of cylinder.

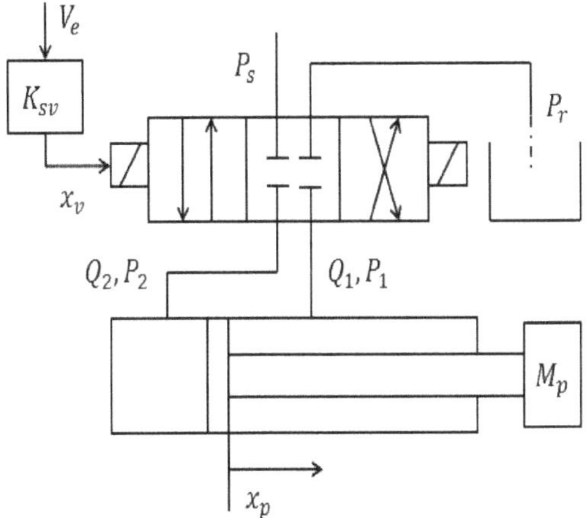

Figure A1. Schematic of Hydraulic Actuator.

The fluid supplied flow rate of the forward chamber and return flow rate of the return chamber are derived by Bernoulli equation respectively and expressed as follow

$$Q_1 = C_d \omega x_v \sqrt{2(P_s - P_1)/\rho} \tag{A1}$$

$$Q_2 = C_d \omega x_v \sqrt{2(P_2)/\rho} \tag{A2}$$

where, x_v is displacement of servo valve spool, C_d is flow coefficient, ω is area gradient of spool valve, ρ is density of fluid,P_s, P_1, P_2 is pressure of supplied fluid and pressure inside the two chambers of the cylinder respectively.

The load pressure P_L and load flow rate Q_L are expressed as follow.

$$P_L = P_1 - P_2 \tag{A3}$$

$$Q_L = (Q_1 + Q_2)/2 \tag{A4}$$

By Using Equations (A3) and (A4), load flow rate of servo value is defined as follow

$$Q_L = \alpha C_d \omega x_v \sqrt{2(P_s - P_L)/\rho} \tag{A5}$$

where

$$\alpha = \frac{1+\eta}{\sqrt{2(1+\eta^2)}} \leq 1 \tag{A6}$$

Furthermore, the flux inside of hydraulic cylinder is defined as Equation (A7) by continuity equation.

$$Q_L = A_e \dot{x}_p + \frac{V_e}{4\beta} \dot{P}_L \tag{A7}$$

where, A_e is average area between piston and piston load, β is the effective flow coefficient modulus and V_e is the volume of the chamber of the cylinder.

The dynamics equation of load system is obtained as

$$A_e P_L = M_p \ddot{x}_p + B_p \dot{x}_p + F_p \tag{A8}$$

where M_p is the total mass elements, B_p is the total damper elements between rod and cylinder, F_p is the total friction elements and x_p is the displacement of rod. To calculate the linearized load flow dynamic equation, Equation (A5) can be linearized as follow

$$Q_L = k_q x_v - k_p P_L \tag{A9}$$

where k_q is load flow coefficient, k_p is load flow pressure coefficient and defined as follow

$$k_q = \alpha C_d \omega x_v \sqrt{(P_s - P_L{}^*)/\rho} = \frac{\partial Q_L}{\partial x_v}|_{P_L = P_L{}^*} \tag{A10}$$

$$k_p = \frac{C_d \omega x_v{}^*}{2}\sqrt{1/\rho(P_A - P_L{}^*)} = \frac{\partial Q_L}{\partial P_L}|_{x_v = x_v{}^*} \tag{A11}$$

$x_v{}^*$ and $P_L{}^*$ in Equations (A10), (A11) are x_v and P_L near the operating point. Using Equation (A7), linearized load flow dynamic equation can be rewritten as Equation (A12). The effective bulk modulus of the fluid is much larger value than other constant parameters. There the pressure variation terms of Equation (A7) is negligible.

$$A_e \dot{x}_p = k_q x_v - k_p P_L \tag{A12}$$

Combing Equations (A7) and (A12), 2nd order dynamic equation can be obtained as follow

$$M_p \ddot{x}_p + (B_p + \frac{A_e^2}{k_p})\dot{x}_p + F_e = \frac{A_e k_q}{k_p} x_v \tag{A13}$$

After separating the non-linear term and parameter error from Equation (A13), the equation of motion is derived as

$$M_{HT}\ddot{x}_p + B_{TH}\dot{x}_p + \psi = KK_{sv}V_e \tag{A14}$$

where M_{HT} and B_{TH} is the equivalent value of mass element and damper element respectively, ψ is the summation of non-linear term, parameter error, friction and disturbance and K is the linear coefficient of $\frac{A_e k_q}{k_p}$. The perturbation ψ, is defined as follow

$$\psi = M_{HT}^{-1}[\Delta K \Delta K_{sv} u_H - \{\Delta M_{HT}\ddot{x}_p + \Delta B_{TH}\dot{x}_p + F_e\}] \tag{A15}$$

References

1. Fisher, J.J. *Applying Robots in Nuclear Applications*; Society of Manufacturing Engineers: Dearborn, MI, USA, 1985.
2. Clement, G.; Vertut, J.; Cregut, A.; Antione, P.; Guittet, J. Remote handling and transfer techniques in dismantling strategy. In Proceedings of the Seminar on Remote Handling in Nuclear Facilities, Oxford, UK, 2–5 October 1984; pp. 556–569.
3. Ma, S.; Hirose, S.; Yoshinada, H. Development of a hyper-redundant multijoint manipulator for maintenance of nuclear reactors. *Adv. Robot.* **1994**, *9*, 281–300. [CrossRef]
4. Denmeade, T. A pioneer's journey into the sarcophagus. *Nucl. Eng. Int.* **1998**, *43*, 18–20.
5. Varley, J. Windscale: Getting down to the core. *Nucl. Eng. Int.* **1997**, *12*, 26–28.
6. Benest, T. Taking up arms for decommissioning. *Nucl. Eng. Int.* **2004**, *49*, 14–16.
7. Tachibana, M.; Shimada, T.; Yanagihara, S. Development of remote dismantling systems for decommissioning of nuclear facilities. In Proceedings of the WM'00 Conference, Tucson, AZ, USA, 27 February–3 March 2000.
8. Bakari, M.J.; Zied, K.M.; Seward, D.W. Development of a multi-arm mobile robot for nuclear decommissioning tasks. *Int. J. Adv. Robot. Syst.* **2007**, *4*, 51. [CrossRef]
9. Dubus, G.; David, O.; Measson, Y.; Friconneau, J.-P.; Palmer, J. Making hydraulic manipulators cleaner and safer: From oil to demineralized water hydraulics. In Proceedings of the IROS 2008. IEEE/RSJ International Conference on Intelligent Robots and Systems, Nice, France, 22–26 September 2008; IEEE: Piscataway, NJ, USA, 2008; pp. 430–437.

10. Chabal, C.; Proietti, R.; Mante, J.-F.; Idasiak, J.-M. Virtual reality technologies: A way to verify dismantling operations, first application case in a highly radioactive cell. In Proceedings of the ACHI, Digital World, Le Gosier, France, 23–28 February 2011.
11. Raju, G.J.; Verghese, G.C.; Sheridan, T.B. Design issues in 2-port network models of bilateral remote manipulation. In Proceedings of the 1989 IEEE International Conference on Robotics and Automation, Scottsdale, AZ, USA, 14–19 May 1989; IEEE: Piscataway, NJ, USA, 1989; pp. 1316–1321.
12. Hashtrudi-Zaad, K.; Salcudean, S.E. Transparency in time-delayed systems and the effect of local force feedback for transparent teleoperation. *IEEE Trans. Robot. Autom.* **2002**, *18*, 108–114. [CrossRef]
13. Lawrence, D.A. Stability and transparency in bilateral teleoperation. *IEEE Trans. Robot. Autom.* **1993**, *9*, 624–637. [CrossRef]
14. Yokokohji, Y.; Yoshikawa, T. Bilateral control of master-slave manipulators for ideal kinesthetic coupling-formulation and experiment. *IEEE Trans. Robot. Autom.* **1994**, *10*, 605–620. [CrossRef]
15. Hannaford, B. A design framework for teleoperators with kinesthetic feedback. *IEEE Trans. Robot. Autom.* **1989**, *5*, 426–434. [CrossRef]
16. Adams, R.J.; Hannaford, B. Stable haptic interaction with virtual environments. *IEEE Trans. Robot. Autom.* **1999**, *15*, 465–474. [CrossRef]
17. Yang, Y.; Hua, C.; Li, J.; Guan, X. Finite-time output-feedback synchronization control for bilateral teleoperation system via neural networks. *Inf. Sci.* **2017**, *406*, 216–233. [CrossRef]
18. Truong, D.Q.; Truong, B.N.M.; Trung, N.T.; Nahian, S.A.; Ahn, K.K. Force reflecting joystick control for applications to bilateral teleoperation in construction machinery. *Int. J. Precis. Eng. Manuf.* **2017**, *18*, 301–315. [CrossRef]
19. Peñaloza-Mejía, O.; Márquez-Martínez, L.A.; Alvarez-Gallegos, J.; Alvarez, J. Master-slave teleoperation of underactuated mechanical systems with communication delays. *Int. J. Control Autom. Syst.* **2017**, *15*, 827–836. [CrossRef]
20. Mellah, R.; Guermah, S.; Toumi, R. Adaptive control of bilateral teleoperation system with compensatory neural-fuzzy controllers. *Int. J. Control Autom. Syst.* **2017**, *15*, 1949–1959. [CrossRef]
21. Farooq, U.; Gu, J.; El-Hawary, M.; Asad, M.U.; Abbas, G. Fuzzy model based bilateral control design of nonlinear tele-operation system using method of state convergence. *IEEE Access* **2016**, *4*, 4119–4135. [CrossRef]
22. Liu, Y.-C.; Khong, M.-H.; Ou, T.-W. Nonlinear bilateral teleoperators with non-collocated remote controller over delayed network. *Mechatronics* **2017**, *45*, 25–36. [CrossRef]
23. Su, X. Master–slave control for active suspension systems with hydraulic actuator dynamics. *IEEE Access* **2017**, *5*, 3612–3621. [CrossRef]
24. Kallu, K.D.; Jie, W.; Lee, M.C. Sensorless reaction force estimation of the end effector of a dual-arm robot manipulator using sliding mode control with a sliding perturbation observer. *Int. J. Control Autom. Syst.* **2018**, *16*, 1367–1378. [CrossRef]
25. Abut, T.; Soyguder, S. Real-time control of bilateral teleoperation system with adaptive computed torque method. *Ind. Robot Int. J.* **2017**, *44*, 299–311. [CrossRef]
26. Islam, S. State and impedance reflection based control interface for bilateral telerobotic system with asymmetric delay. *J. Intell. Robot. Syst.* **2017**, *87*, 425–438. [CrossRef]
27. Le, M.-Q. *Development of Bilateral Control for Pneumatic Actuated Teleoperation System*; INSA: Lyon, France, 2011.
28. Patil, M.D.; Abukhalil, T. Design and implementation of heterogeneous robot swarm. In Proceedings of the ASEE 2014 Zone I Conference, Bridgpeort, CT, USA, 3–5 April 2014; pp. 3–5.
29. Xu, X.; Cizmeci, B.; Schuwerk, C.; Steinbach, E. Model-mediated teleoperation: Toward stable and transparent teleoperation systems. *IEEE Access* **2016**, *4*, 425–449. [CrossRef]
30. Sun, D.; Naghdy, F.; Du, H. Time domain passivity control of time-delayed bilateral telerobotics with prescribed performance. *Nonlinear Dyn.* **2017**, *87*, 1253–1270. [CrossRef]
31. Azimifar, F.; Abrishamkar, M.; Farzaneh, B.; Sarhan, A.A.D.; Amini, H. Improving teleoperation system performance in the presence of estimated external force. *Robot. Comput.-Integr. Manuf.* **2017**, *46*, 86–93. [CrossRef]
32. Sakaino, S.; Furuya, T.; Tsuji, T. Bilateral control between electric and hydraulic actuators using linearization of hydraulic actuators. *IEEE Trans. Ind. Electron.* **2017**, *64*, 4631–4641. [CrossRef]

33. Soltani, M.K.; Khanmohammadi, S.; Ghalichi, F.; Janabi-Sharifi, F. A soft robotics nonlinear hybrid position/force control for tendon driven catheters. *Int. J. Control Autom. Syst.* **2017**, *15*, 54–63. [CrossRef]
34. Wang, Y.; Jiang, S.; Chen, B.; Wu, H. Trajectory tracking control of underwater vehicle-manipulator system using discrete time delay estimation. *IEEE Access* **2017**, *5*, 7435–7443. [CrossRef]
35. Liu, X.; Jiang, W.; Dong, X.-C. Nonlinear adaptive control for dynamic and dead-zone uncertainties in robotic systems. *Int. J. Control Autom. Syst.* **2017**, *15*, 875–882. [CrossRef]
36. Li, Y.F.; Chen, X.B. On the dynamic behavior of a force/torque sensor for robots. *IEEE Trans. Instrum. Measur.* **1998**, *47*, 304–308. [CrossRef]
37. Katsura, S.; Matsumoto, Y.; Ohnishi, K. Analysis and experimental validation of force bandwidth for force control. *IEEE Trans. Ind. Electron.* **2006**, *53*, 922–928. [CrossRef]
38. Wang, J.; Dad, K.; Lee, M.C. Bilateral control of hydraulic servo system based on smcspo for 1dof master slave manipulator. In Proceedings of the 2017 International Conference on Artificial Lifeand Robotics (ICAROB2017), Miyazaki, Japan, 19–22 January 2017.
39. Moura, J.T.; Elmali, H.; Olgac, N. Sliding mode control with sliding perturbation observer. *J. Dyn. Syst. Measur. Control* **1997**, *119*, 657–665. [CrossRef]
40. Slotine, J.J.; Sastry, S.S. Tracking control of non-linear systems using sliding surfaces with application to robot manipulators. In Proceedings of the 1983 American Control Conference, San Francisco, CA, USA, 22–24 June 1983; pp. 132–135.
41. Moura, J.T.; Ghosh Roy, R.; Olgac, N. Sliding mode control with perturbation estimation (smcpe) and frequency shaped sliding surfaces. *J. Dyn. Syst. Measur. Control* **1997**, *119*, 584–588. [CrossRef]
42. Slotine, J.-J.E.; Li, W. *Applied Nonlinear Control*; Prentice Hall: Englewood Cliffs, NJ, USA, 1991; Volume 199.
43. Cho, T.-D.; Seo, S.-H.; Yang, S.-M. A study on the robust position control of single-rod hydraulic system. *J. Korean Soc. Precis. Eng.* **1999**, *16*, 128–135.
44. Jelali, M.; Kroll, A. *Hydraulic Servo-Systems: Modelling, Identification and Control*; Springer Science & Business Media: Berlin, Germany, 2012.
45. Lee, M.C.; Aoshima, N. Identification and its evaluation of the system with a nonlinear element by signal compression method. *Trans. Soc. Instrum. Control Eng.* **1989**, *25*, 729–736. [CrossRef]

Article

Time Sequential Motion-to-Photon Latency Measurement System for Virtual Reality Head-Mounted Displays

Song-Woo Choi, Siyeong Lee, Min-Woo Seo and Suk-Ju Kang *

Department of Electronic Engineering, Sogang University, Seoul 04107, Korea; songwoo602@gmail.com (S.-W.C.); siyeong@sogang.ac.kr (S.L.); yoynok08@gmail.com (M.-W.S.)

* Correspondence: sjkang@sogang.ac.kr; Tel.: +82-02-705-8466

Received: 12 August 2018; Accepted: 29 August 2018; Published: 1 September 2018

Abstract: Because the interest in virtual reality (VR) has increased recently, studies on head-mounted displays (HMDs) have been actively conducted. However, HMD causes motion sickness and dizziness to the user, who is most affected by motion-to-photon latency. Therefore, equipment for measuring and quantifying this occurrence is very necessary. This paper proposes a novel system to measure and visualize the time sequential motion-to-photon latency in real time for HMDs. Conventional motion-to-photon latency measurement methods can measure the latency only at the beginning of the physical motion. On the other hand, the proposed method can measure the latency in real time at every input time. Specifically, it generates the rotation data with intensity levels of pixels on the measurement area, and it can obtain the motion-to-photon latency data in all temporal ranges. Concurrently, encoders measure the actual motion from a motion generator designed to control the actual posture of the HMD device. The proposed system conducts a comparison between two motions from encoders and the output image on a display. Finally, it calculates the motion-to-photon latency for all time points. The experiment shows that the latency increases from a minimum of 46.55 ms to a maximum of 154.63 ms according to the workload levels.

Keywords: head-mounted display; virtual reality; motion-to-photon latency

---

1. Introduction

Because the personal computer's performance has been greatly improved, real-time rendering of high-quality images has become possible, and virtual reality (VR) technology has become a reality [1]. According to this trend, a variety of VR devices utilizing 3D rendering and sensor-based technology have been released. The head-mounted display (HMD) device, which is a system designed to improve immersion by mounting a wide field-of-view display within the user's sight, has been gaining popularity [2]. However, because of the mismatch between visual and vestibular systems, users wearing HMD devices experience motion sickness and dizziness, which can be an obstacle to the VR market. The motion-to-photon latency, which is one of the causes for this mismatch, is the time delay for a user movement to be fully reflected on a display screen [3]. Generally, three steps are required to render an image, as shown in Figure 1. When head motion occurs, the motion detection unit samples the orientation data for the view generation. After the motion detection, the visual processing unit renders a 3D image. Finally, the rendered image is outputted to a display corresponding to the head orientation of the user measured by the sensor. As these steps take time, the delay results in motion-to-photon latency. In this case, the image does not exactly correspond to the actual head orientation of the user, thereby causing the user to experience motion sickness [4]. Therefore, many HMD makers such as Oculus VR have conducted studies to minimize the mismatch caused by the motion-to-photon latency.

Numerous studies such as prediction techniques based on the data acquired by inertial measurement unit (IMU) sensors [5] and asynchronous time warp (ATW) [6] have been conducted to overcome this limitation. The quantitative evaluation of the above methods required the authors to measure the motion-to-photon latency. They could then improve and evaluate these methods with reference data.

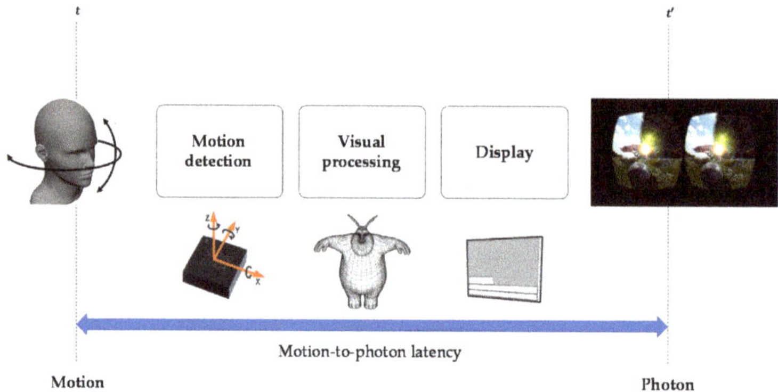

Figure 1. Image generation process in the HMD (head-mounted display) system.

To measure this motion-to-photon latency, the method of [7] presents a low-cost system with high accuracy which measures the latency. However, it can be used for the mobile device. The method of [8] proposes the measurement system and a technique to reduce the latency for the optical see-through display. The authors had previously proposed a new measurement system by using multiple sensors such as an optical sensor and encoders and by comparing the physical signal to the luminance signal from a VR scene reflected on the display [9]. Figure 2 shows the overall architecture of this measurement system. A rotary platform, which controls the physical motion, was proposed in that method to simulate the rotation of the neck. It measured the change of the image brightness when the physical motion commenced. The method could directly measure the motion-to-photon latency because it calculated the time difference between the point at which the brightness startedchanging and that at which the physical motion started However, although this method provided accurate measurement results, its limitation was that the latency could only be measured when the physical motion began.

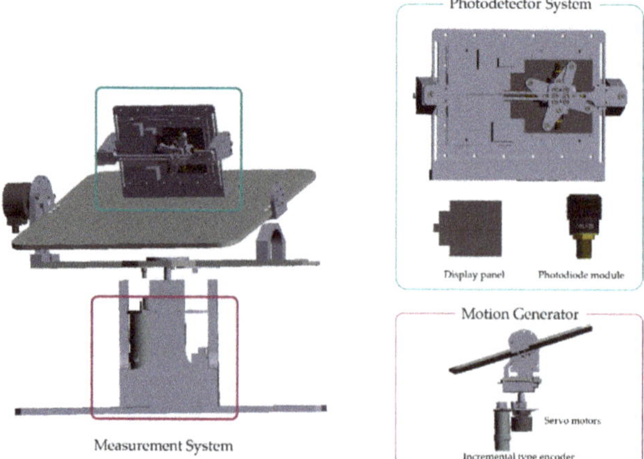

Figure 2. Overall architecture of the previously proposed motion-to-photon latency measurement system.

In this paper, the authors propose a novel time sequential measurement system for the motion-to-photon latency by comparing the physical motion and the motion reflected on the image by 3D rendering. Therefore, it is possible to measure and record the time sequential latency more precisely in real-time, while reflecting the change in workload over time, unlike the conventional method mentioned in [9]. Specifically, the proposed method renders the rotation data with the intensity levels of pixels on the measurement area, which is specially invented for the HMD system, and it can obtain the motion-to-photon latency data in all temporal ranges. Therefore, it has a great advantage over the existing method.

This paper is organized as follows. Section 2 describes the proposed latency measurement system with five subsections. Section 3 presents the experimental environment and the results using the proposed system. Section 4 presents the paper's conclusion.

2. Proposed Method

Figure 3 shows the overall block diagram of the proposed system. The proposed method measured the actual angular change of the encoder when physical movement occurred. Concurrently, when the physical movement was measured by the IMU sensor, the rendered image was outputted to the display reflecting the measured angle, and the corresponding angle value was converted into the intensity image. The proposed method compared these two values. Specifically, a measurement area, which was constructed with multiple 2D objects mapped with an intensity level converting the rotation angle measured from the IMU, was implemented. Then, the photodetector converted the luminance reflected in the measurement area into a voltage. Finally, the oscilloscope measured this voltage value to determine the intensity level. At the same time, the pulse obtained from the encoder was converted into a position value and compared with the luminance-based position value obtained from the photodetector in real time. Finally, the motion-to-photon latency was calculated by measuring the time difference at which the angle measured from the display reached the same rotation angle of the encoder, as shown in Figure 3. The detailed processes are explained in following sections.

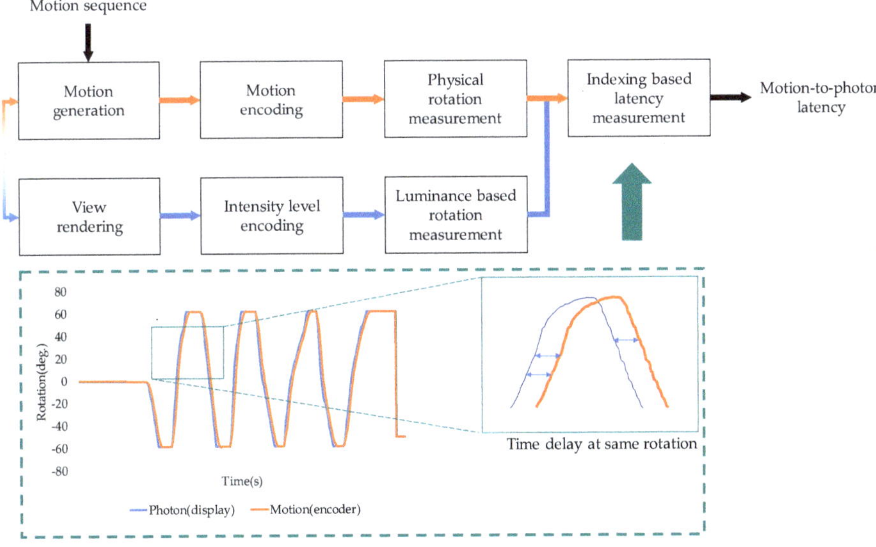

Figure 3. Overall block diagram of the proposed system.

2.1. Motion Generator-Based on Human Neck Movement

The authors used the rotary platform proposed in [9] to model the physical rotation of the user's neck. Specifically, in order to simulate a human neck, a two-degree-of-freedom rotary platform was proposed, built with joints and links based on head kinematics. Therefore, the motion at the end-effector could be estimated through the motion in each joint based on forward kinematics [10,11], as shown in Figure 4.

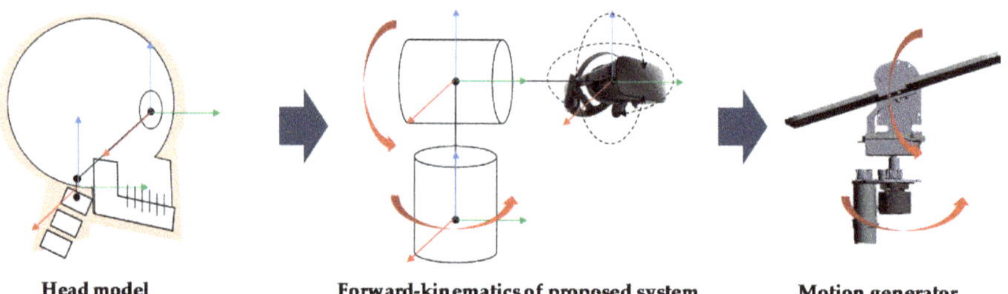

Figure 4. Kinematics of the head-model-based measurement platform.

For this paper, the authors implemented the kinematic analysis of the model using the Denavit–Hartenberg parameters [12], and the angle of the end-effector was estimated by calculating the angle of each joint [13]. In the proposed method, the rotation of the end-effector was estimated by performing rotation transformation on the displacement and rotation in the previous link, and coordinate transformation with the translation matrix. The transformation was carried out on n number of links with the following matrix multiplication:

$$[T] = [Z_1][X_1][Z_2][X_2]\cdots[X_n][Z_n], \tag{1}$$

where $[T]$ denotes the transformation matrix of the end-effector, $[Z_n]$ denotes a transformation matrix at the n-th joint, and $[X_n]$ denotes a transformation matrix of at the n-th link. The transformation matrix of the i-th joint is as follows:

$$[Z_i] = \begin{bmatrix} \cos\theta_i & -\sin\theta_i & 0 & 0 \\ \sin\theta_i & \cos\theta_i & 0 & 0 \\ 0 & 0 & 1 & d_i \\ 0 & 0 & 0 & 1 \end{bmatrix}, \tag{2}$$

where θ_i denotes the rotation angle between the previous joint and the next i-th joint along the z axis, and d_i denotes the displacement between each joint. The transformation matrix of the i-th link is as follows:

$$[X_i] = \begin{bmatrix} 1 & 0 & 0 & r_{i,i+1} \\ 0 & \cos\alpha_{i,i+1} & -\sin\alpha_{i,i+1} & 0 \\ 0 & \sin\alpha_{i,i+1} & \cos\alpha_{i,i+1} & 0 \\ 0 & 0 & 0 & 1 \end{bmatrix}, \tag{3}$$

where $\alpha_{i,i+1}$ denotes an angle between each link along the x-axis, and $r_{i,i+1}$ denotes the displacement between each link. Finally, the transformation matrix for the forward kinematics from the $(n-1)$-th link to the n-th link is as follows:

$$T_n = \begin{bmatrix} \cos\theta_n & -\sin\theta_n\cos\alpha_n & \sin\theta_n\sin\alpha_n & r_n\cos\theta_n \\ \sin\theta_n & \cos\theta_n\cos\alpha_n & -\cos\theta_n\sin\alpha_n & r_n\sin\theta_n \\ 0 & \sin\alpha_n & \cos\alpha_n & d_n \\ 0 & 0 & 0 & 1 \end{bmatrix}. \tag{4}$$

Therefore, the Euler angles [14] of the end-effector were calculated from the inner rotation matrix. The Euler angles were used in an interface of the proposed system.

2.2. Measurement Area Design

In the proposed method, photodetectors measured the rotation angles from the rendered image. The advantage of photodetectors is that they can be measured more accurately while modeling the human visual function owing to their fast response time. However, the photodetectors only measure the intensity of light within the specific range. Hence, the authors proposed specially invented objects that displayed the current rotation angle on a VR scene. Specifically, the objects did not have any colors but had grayscale intensities, and the intensity levels represented the rotation angle with the pre-defined conversion rule that converted the rotation angle with a float type into the intensity value with an unsigned integer type. Therefore, by measuring the luminance for the objects in the measurement area, the current rotation angle was obtained.

When designing the objects in the measurement area, the shapes of the object could influence not only the rendering performance but also the measurement performance. In 3D rendering, there is back-face culling in that the game engine does not calculate the region occluded by an object [15]. Therefore, due to occlusion in the measurement area, the rendering performance is changed. Since the back-face culling affects the latency measurement, the size of the measurement area to be projected into the display is optimally adjusted to minimize the back-face culling and to maximize the voltage that the active layer of the photodetector can measure.

The brightness of the measurement area was determined by the rotation value calculated from the IMU sensor of the HMD device when rendering the current frame. To enhance the measurement accuracy, two objects represented the current rotation angle at a single rotation axis. Arrangement of the four objects enabled both the yaw and pitch, which were the target rotation directions, to be calculated.

The scanline of the display for the area to be measured should also be considered. Since the entire image is not drawn at once but sequentially drawn along the scanline, there is a time difference depending on the arrangement of the measurement area [16]. In order to estimate the correct angle, two measurement areas must be synchronized considering the physical location.

2.3. Voltage Mapping Table

When a change in light is detected in the photodetector installed in the measurement area, the current (the output voltage as a result) emitted by the internal photodiode changes. If the experimental environment and parameters are controlled constantly, the output voltage of the photodetector to the image luminance is constant. This means that the output voltage can be converted back to the original luminance information. Therefore, if the conversion between the luminance and the voltage is known in advance, the rotation angle applied in the rendering process can be estimated. In the proposed method, the relationship between the voltage and luminance was acquired in advance, and the mapping table for the relationship was obtained by using these data. Figure 5 shows how to estimate intensity levels from the luminance of the measurement area on the display using the mapping table.

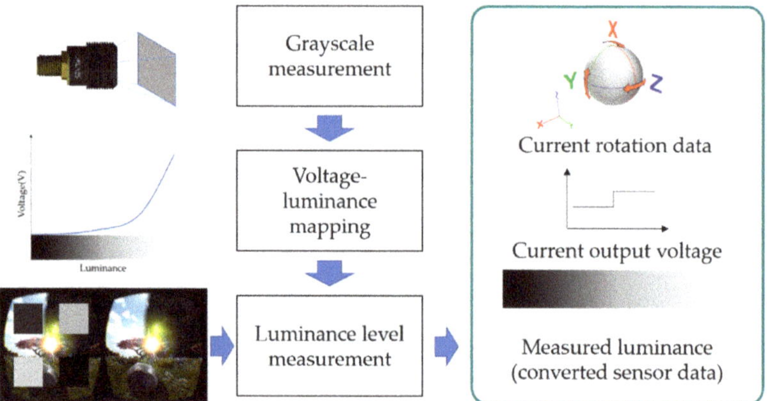

Figure 5. Luminance estimation from an object rendered by using the photodetector system and the voltage mapping table.

The voltage of the photodetector is proportional to the luminance, and therefore, the low luminance has a low output voltage. In this case, the amount of the voltage according to the brightness was not distinguishable from the noise when the voltage outputted from the measurement region was low at the stage of constructing the mapping table. Figure 6 illustrates the above problem. Therefore, low voltage levels are excluded from the measurement, and the mapping table is estimated from the upper level. Then, the current rotation value was inferred. There is a problem that if only the 128 levels of intensity are used for 256 gray levels with an 8-bit depth, the measurement resolution is halved. To compensate for this problem, an additional photodetector was used. An additional sensor restored the resolution to the original resolution by using double photodetectors to measure one rotation of the axis.

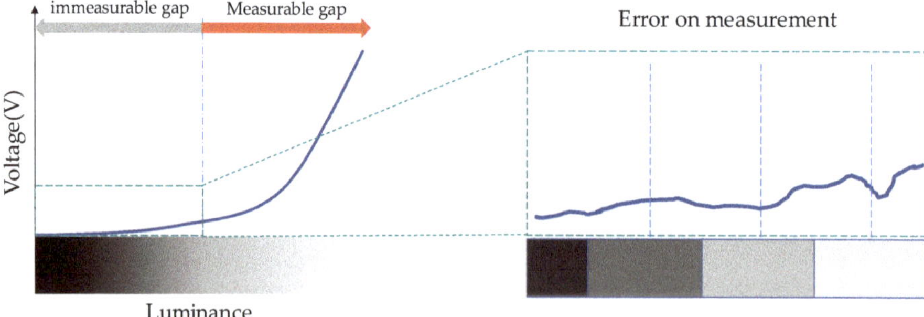

Figure 6. Measurement error and noise for an output voltage in the photodetector system.

For double photodetectors, a new formula was needed which converted a rotation value into two intensity levels. The rotation angle was split into two intensity levels, and these intensity levels were reflected on the measurement area so that the photodetectors measured each intensity level to convert them back to the rotation angle. The specific conversion formula is as follows:

$$\text{Channel} = \left\{ \begin{array}{l} MSBs_{7bit}\left[\frac{(\theta+M_{euler})}{2M_{euler}} \times 2^{14}\right] + 127 \\ LSBs_{7bit}\left[\frac{(\theta+M_{euler})}{2M_{euler}} \times 2^{14}\right] + 127 \end{array} \right\}, \text{ for } |\theta| \le M_{euler}, \quad (5)$$

where M_{euler} denotes the maximum Euler angle in the measurement system, θ denotes the current angle measured from HMD, and $MSBs_{7bit}$ and $LSBs_{7bit}$ denote the upper and lower 7 bits of the converted value, respectively. In the measurement area, 14-bit data were mapped onto the objects.

For example, as shown in Figure 7, the current rotation of the yaw axis and maximum Euler angle are given. Then, these values were calculated with a conversion formula. The current rotation angle was converted into binary data and split into two parts. Each part was the intensity level of the measurement area. After estimating these intensity levels by the photodetectors, the reverse calculation was performed with the inverse of the conversion formula. Finally, the current rotation was obtained, which is used for the generation of a VR scene.

Figure 7. Example of the intensity level conversion and its inverse for measuring the current rotation angle.

However, in some cases, accurate mapping tables could not be formed because of changes in the external environment during the measurement process and changes in the output voltage due to the electrical noise. To solve this problem, a polynomial regression to store the data was performed from the voltage data obtained by repeated experiments. In the proposed method, an optimal curve was obtained by repeatedly performing a polynomial regression of less than third-order to avoid overfitting. Then, the obtained curve was used for the voltage estimation for a luminance level.

The third polynomial regression used in this paper is described below. The regression model is as follows:

$$v_i = a_0 + a_1 I_1 + a_2 I_2^2 + a_3 I_3^3 + \varepsilon_i \ (i = 1, 2, 3, \cdots, n), \tag{6}$$

where n denotes the number of points, v_i denotes an output voltage from the photodetector, I_i denotes an 8-bit unsigned integer value of the intensity level projected on the measurement area, a_i denotes a curvature parameter, and ε_i denotes a random error. The polynomial curve is composed as follows.

$$\begin{bmatrix} v_1 \\ v_2 \\ v_3 \end{bmatrix} = \begin{bmatrix} 1 & I_1 & I_1^2 & I_1^3 \\ 1 & I_2 & I_2^2 & I_2^3 \\ 1 & I_3 & I_3^2 & I_3^3 \end{bmatrix} \begin{bmatrix} a_1 \\ a_2 \\ a_3 \end{bmatrix} + \begin{bmatrix} \varepsilon_1 \\ \varepsilon_2 \\ \varepsilon_3 \end{bmatrix}. \tag{7}$$

In (7), if the random error term is excluded, the curvature parameters can be estimated after the least-squares estimation [17]. The least-squares estimation of these parameters is as follows:

$$\boldsymbol{a} = \left(\boldsymbol{I}^T \boldsymbol{I}\right)^{-1} \boldsymbol{I}^T \boldsymbol{v}, \tag{8}$$

where $\boldsymbol{v}$ denotes a vector composed of the output voltages from the photodetector, $\boldsymbol{a}$ denotes the curvature parameter vector of the voltage–luminance curve, and I denotes the matrix of intensity levels from the repeated measurements. Using this regression model, the motion-to-photon latency was measured and is described in the following section.

2.4. Measurement of Motion-to-Photon Latency

The output image on the display was rendered by rotating the HMD using a motion generator designed to control the rotation of the HMD device. It controlled the HMD device with the desired rotation angle using a highly accurate DC servo motor. An internal encoder in the motor, which was used to control the motor, could also be used for measuring the current physical movement. However, in order to more precisely measure the physical movement over time, the incremental type encoders, additionally mounted on the axes, detected the rotation angles and outputted them as pulses. Concurrently, by measuring the luminance of the objects, the current angle data could be obtained by the conversion rule with the mapping table described above. The motion data about the current frame outputted from the display and the motion data outputted from the motion generator were compared. Due to the motion-to-photon latency, the rotation angle from the display was temporally lagged with respect to the angle outputted from the motion generator. To quantify the motion-to-photon latency, the proposed system calculated the time at which the angle, outputted from the display, matched the angle outputted from encoders of the motion generator. Finally, it calculated the difference, which was the motion-to-photon latency between the two time points that had the same rotation angle. When calculating the time points, the authors found the indices of two rotation buffers with the same rotation angles, as shown in Figure 8. To avoid finding multiple points of the same value, the searching range was limited. There were multiple points that had the same value in all ranges. By limiting the searching range, the method allowed the buffers to be searched in the limited range so that only one same value was selected.

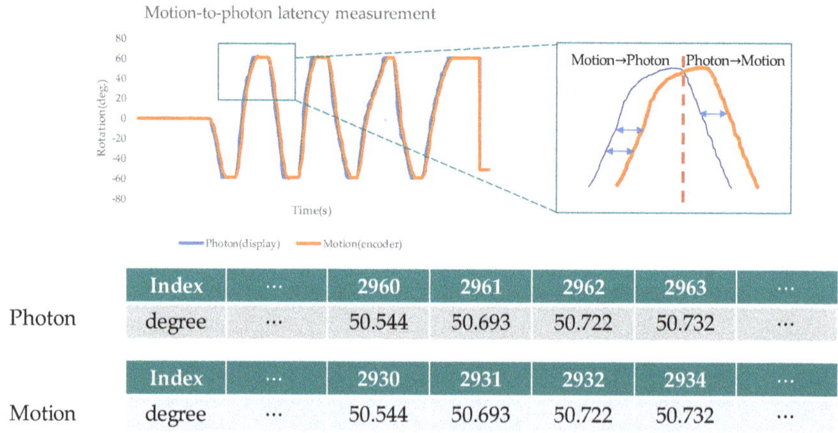

	Index	···	2960	2961	2962	2963	···
Photon	degree	···	50.544	50.693	50.722	50.732	···

	Index	···	2930	2931	2932	2934	···
Motion	degree	···	50.544	50.693	50.722	50.732	···

Figure 8. Example of the motion-to-photon latency calculation.

2.5. Real-Time Measurement and Interface

The proposed method used buffers to store the data from the encoders and the display. Specifically, it synchronized and stored all data in an oscilloscope, and the time delays were compared by searching points that had the same values. Once each time difference was found, it was converted to a time value to obtain the final motion-to-photon latency. In addition, a user interface was proposed to control the system and to integrate and run several modules. Figure 9 shows the configuration of the interface. The sequence-based motion generation part provided the information for controlling the motion generator. In the data plotting part, the graphs for the rotation angles are plotted. The latency measurement part provides the current, mean, and max values for the motion-to-photon latency.

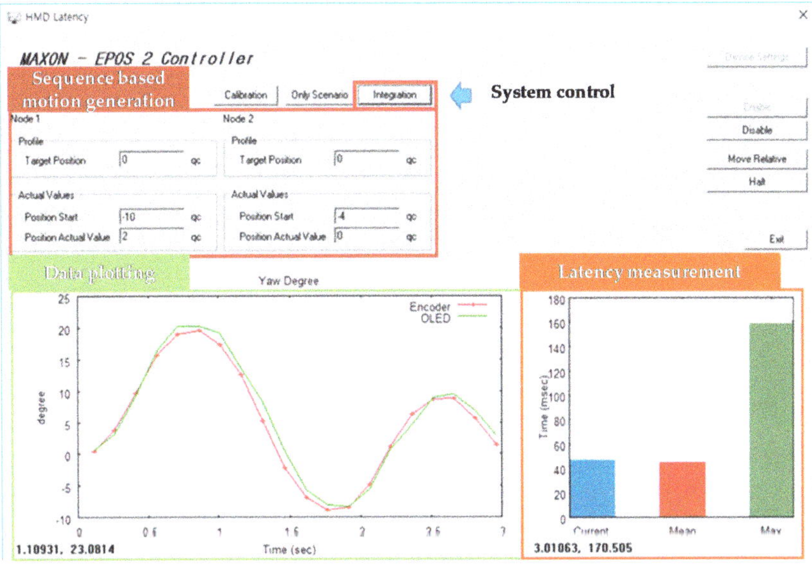

Figure 9. Proposed user interface of the motion-to-photon latency measurement system.

3. Experimental Results

The experimental environment is as follows. First, the authors used an Oculus Rift DK 2 HMD device (Oculus VR, Menlo park, CA, USA) for the VR implementation [18]. To drive a motion generator, two RE 40 (Maxon Motor, Sachseln, Switzerland) [19] were used as a DC motor along two axes, and the same number of EPOS2 50/5 (Maxon Motor, Sachseln, Switzerland) were used as controllers to control them [20]. However, since the position estimated from the motor itself had mechanical latency, additional encoders (EIL 580 Baumer, Frauenfeld, Switzerland), which are of the incremental type with 500 steps/turn and 300 kHz frequency, were used to measure more accurate current rotation angles [21]. SM05PDs (THORLABS, Newton, NJ, USA) were used as the photodetectors, which have a spectral range from 200 nm to 1000 nm [22]. A PicoScope 4824 (Pico Technology, Saint Neots, UK) was used to measure the voltages from encoders and photodetectors [23]. It measured the voltages of the encoder and photodetector with a sampling rate of 100,000 times per second and a bandwidth of 5.0 Gbps, thereby allowing nearly continuous data description and processing in real time. To render the VR scenes and drive the entire proposed system, a PC, which had an Intel i7-6700k @ 4.4 GHz CPU [24] and an NVIDIA Geforce GTX 1080 graphic card [25], were used. The experiment was conducted by the following two methods. First, the authors measured the changes in the latency according to the change in the position sequence while the initial workload was fixed. In this case, the workload was assigned the minimum level. The second method used only one of the motion sequences for different initial workload levels. Figure 10 shows the overall experimental process described above.

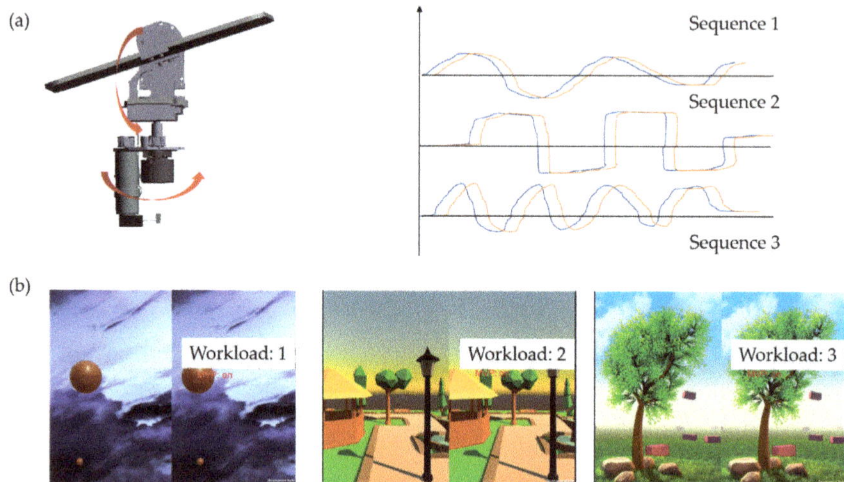

Figure 10. Overall experimental processes for calculating the motion-to-photon latency in (**a**) different sequences and (**b**) different workload levels.

3.1. Position Sequence

Table 1 lists the position sequences defined in the experiment for the yaw and pitch axes. Six peak points were inputted to the motion generator controlled by the input angle, and the system was controlled according to the sequence shown in Figure 11. In this case, as shown in Table 2, the average latency was up to 46.55 ms and the maximum latency was up to 63.72 ms. In this experiment, the positions and shapes of the 3D objects in a VR scene were changed according to the orientation of the HMD. By reflecting these changes of the objects, the rendering workload was changed with time, thereby changing the motion-to-photon latency. As shown in Table 2, the motion-to-photon latency was varied according to the motion sequence. These results show that the proposed system could measure the motion-to-photon latency in all temporal ranges when the position sequence changed.

Table 1. Peak points of simulated head orientations in yaw and pitch directions.

Orientation		**Yaw**		**Pitch**	
Number of Sequences		**1**	**2**	**1**	**2**
Peak point	1	0.00	0.00	0.00	0.00
	2	19.10	13.50	−14.00	19.10
	3	5.10	−14.20	−5.00	5.10
	4	0.00	−15.00	13.50	0.00
	5	−8.90	5.00	6.50	−8.90
	6	0.00	0.00	0.00	0.00

Table 2. Minimum, maximum, and average motion-to-photon latencies by different position sequence.

Orientation	**Yaw**		**Pitch**	
Number of Sequences	**1**	**2**	**1**	**2**
Min. (ms)	21.83	15.78	23.42	32.10
Max. (ms)	63.72	69.33	53.75	65.36
Average (ms)	46.55	47.42	43.87	49.45

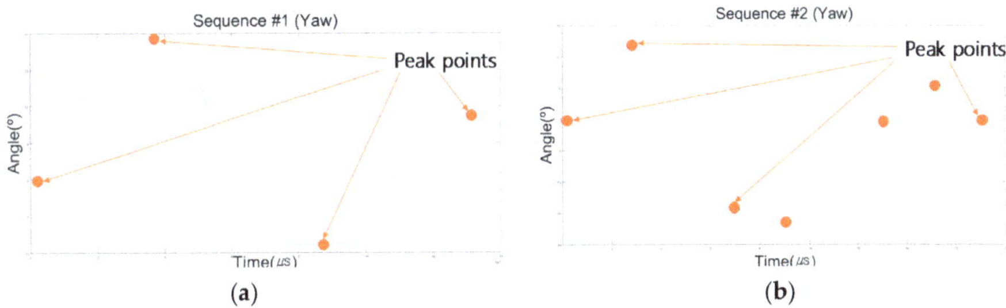

Figure 11. Sequences and peak points for simulated head orientations: the change of angles in (**a**) sequence #1 and (**b**) sequence #2.

3.2. Different Workloads

The authors also measured the latency change while varying the initial rendering workload with the fixed position sequence. The steps of each workload were determined by the number of vertices, which is the minimum unit that constitutes the mesh that has the greatest influence on the computation when rendering the image. In the workload of level 0, the minimum number of objects was two and the number of vertices was 800,000. In the workload of level 3, the number of objects was 17, and the number of vertices was 9.5 million. Table 3 summarizes the number of vertices per level. An increase in the number of vertices also increased the computation time, which directly affected the motion-to-photon latency. In the experiment, the minimum, maximum, and average motion-to-photon latency were measured according to the workload level as shown in Table 4. In this case, the average latency measured in the workload with level 0 was up to 46.55 ms, and the maximum latency was up to 63.75 ms. On the other hand, the average latency measured in the workload with level 3 was up to 154.63 ms, and the maximum latency was up to 198.24 ms. Figure 12 shows the results as a function of the workload level. In the figure, the blue points are the rotation angles from a VR scene, and the orange points are the angles from the motion generator. In the cropped image, the blue points were lagged behind orange points, which are the current angles of the motion. As shown in both of the cropped regions, the time delay of the two points was increased when the workload level was increased. Therefore, when the workload level was changed, the time delay also changed significantly.

Table 3. Various rendering workloads and vertices.

Rendering Workload	0	1	2	3
Number of objects	2	7	12	17
Vertices	0.8 M	2.4 M	4.8 M	9.5 M

Table 4. Minimum, maximum, and average motion-to-photon latencies by different rendering workload levels.

Rendering Workload	0	1	2	3
Min. (ms)	21.80	45.30	88.30	120.15
Max. (ms)	63.75	89.75	133.65	198.24
Average (ms)	46.55	63.22	101.29	154.63

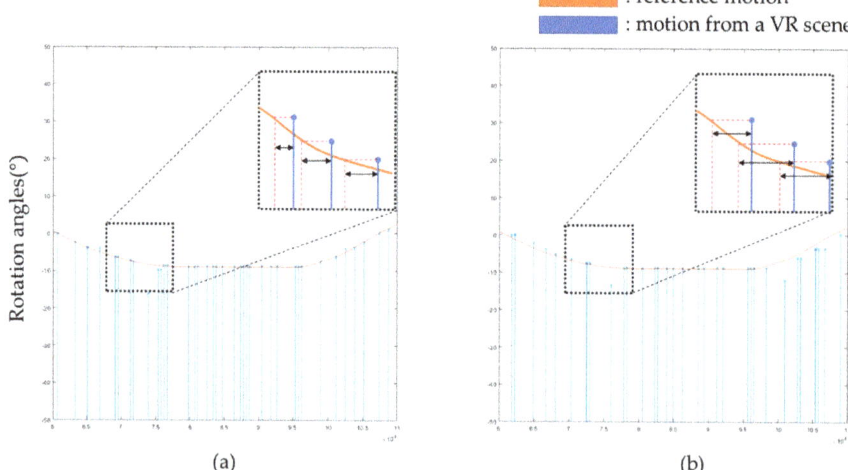

Figure 12. Profiled motion data measured from the physical motion and a virtual reality (VR) output image: (**a**) workload level 0 and (**b**) workload level 3.

4. Conclusions

This paper proposed a novel measurement system to visualize the motion-to-photon latency with time-series data in real time. The proposed system rotated the HMD device using a motion generator based on the kinematics of a human neck to control its motions. In addition, we proposed a measurement area, which includes four objects with intensity levels that are converted from an IMU sensor data of the HMD, to acquire the current motion from the display. Finally, the motion data obtained from the encoder were compared with the motion data obtained from the display to calculate the motion-to-photon latency and visualize it on the dedicated interface. In the experiment, the latency was changed from a minimum of 46.55 ms to a maximum of 154.63 ms according to the workload levels.

Author Contributions: S.-W.C., M.-W.S. and S.-J.K. conceived and designed the experiments; S.-W.C., S.L. and S.-J.K. performed the experiments, and analyzed the data; S.-W.C., M.-W.S., S.L. and S.-J.K. contributed the equipment development; S.-W.C. and S.-J.K. wrote the paper.

Acknowledgments: This research was supported by the National Research Foundation of Korea (NRF) through a grant funded by the Government of Korea (MSIT) (No. 2018R1D1A1B07048421), and supported by the MSIT (Ministry of Science and ICT), Korea, under the ITRC (Information Technology Research Center) support program (IITP-2018-0-01421) supervised by the IITP (Institute for Information & Communications Technology Promotion).

Conflicts of Interest: The authors declare no conflict of interest.

References

1. Chang, S.N.; Chen, W.L. Does visualize industries matter? A technology foresight of global Virtual Reality and Augmented Reality Industry. In Proceedings of the 2017 International Conference on Applied System Innovation (ICASI), Sapporo, Japan, 13–17 May 2017.
2. Azuma, R.; Baillot, Y.; Behringer, R.; Feiner, S. Recent advances in augmented reality. *IEEE Comput. Graph. Appl.* **2001**, *21*, 34–47. [CrossRef]
3. Iribe, B. Virtual reality—A new frontier in computing. In Proceedings of the AMD Announces 2013 Developer Summit, San Jose, CA, USA, 13 November 2013.
4. Justin, M.; Diedrick, M.; Stoffregen, T. The virtual reality head-mounted display Oculus Rift induces motion sickness and is sexist in its effects. *Exp. Brain Res.* **2017**, *3*, 889–901.

5. LaValle, S.M.; Yershova, A.; Katsev, M.; Antonov, M. Head tracking for the Oculus Rift. In Proceedings of the 2014 IEEE International Conference on Robotics and Automation (ICRA), Hong Kong, China, 31 May–7 June 2014.
6. Evangelakos, D.; Mara, M. Extended TimeWarp latency compensation for virtual reality. In Proceedings of the 20th ACM SIGGRAPH Symposium on Interactive 3D Graphics and Games, Redmond, WA, USA, 27–28 February 2016.
7. Tsai, Y.J.; Wang, Y.X.; Ouhyoung, M. Affordable system for measuring motion-to-photon latency of virtual reality in mobile devices. In Proceedings of the SIGGRAPH Asia 2017 Posters, Bangkok, Thailand, 27–30 November 2017.
8. Lincoln, P.; Blate, A.; Singh, M.; Whitted, T.; State, A.; Lastra, A.; Fuchs, H. From motion to photons in 80 microseconds: Towards minimal latency for virtual and augmented reality. *IEEE Trans. Visual Comput. Graph.* **2016**, *22*, 1367–1376. [CrossRef] [PubMed]
9. Seo, M.; Choi, S.; Lee, S.; Lee, E.; Baek, J.; Kang, S. Photosensor-based latency measurement system for head-mounted displays. *Sensors* **2017**, *17*, 1112. [CrossRef] [PubMed]
10. Sharkey, M.; Murray, W.; Heuring, J. On the kinematics of robot heads. *IEEE Trans. Rob. Autom.* **1997**, *13*, 437–442. [CrossRef]
11. Funda, J.; Taylor, H.; Paul, P. On homogeneous transforms, quaternions, and computational efficiency. *IEEE Trans. Rob. Autom.* **1990**, *6*, 382–388.
12. Denavit, J.; Hartenberg, S. A kinematic notation for lower-pair mechanisms based on matrices. *J. Appl. Mech.* **1955**, *1*, 215–221.
13. Kim, J.; Kumar, R. Kinematics of robot manipulators via line transformations. *J. Rob. Syst.* **1990**, *7*, 649–674. [CrossRef]
14. Kucuk, S.; Bingul, Z. Robot kinematics: Forward and inverse kinematics. In *Industrial Robotics: Theory, Modelling and Control*; INTECH Open: London, UK, 2006.
15. Eberly, D.H. *3D Game Engine Design: A Practical Approach to Real-Time Computer Graphics*; CRC Press: Boca Raton, FL, USA, 2006.
16. Jack, K.; Tsatsulin, V. *Dictionary of Video and Television Technology*; Gulf Professional Publishing: Houston, TX, USA, 2002.
17. Kariya, T.; Kurata, H. Generalized least squares estimators. In *Generalized Least Squares*, 1st ed.; John Wiley & Sons: Hoboken, NJ, USA, 2004; pp. 25–67.
18. Oculus Rift DK 2 Overview of the DK2 and SDK 0.4. Available online: https://developer.oculus.com/documentation/pcsdk/0.4/concepts/dg-intro-version/ (accessed on 2 September 2017).
19. Maxon RE 40 40 mm, Graphite Brushes, 150 Watt Specifications. Available online: https://www.maxonmotor.com/medias/sys_master/root/8825409404958/17-EN-132.pdf (accessed on 2 September 2017).
20. EPOS2 50/5, Digital Positioning Controller, 5 A, 11 - 50 VDC Specifications. Available online: https://www.maxonmotor.com/medias/sys_master/root/8821075705886/16-421-422-423-425-EN.pdf (accessed on 2 September 2017).
21. EIL580 Mounted Optical Incremental Encoders Specification. Available online: http://www.baumer.com/es-en/products/rotary-encoders-angle-measurement/incremental-encoders/58-mm-design/eil580-standard (accessed on 2 September 2017).
22. SM05PD Mounted Photodiodes – SM05 and SM1 Compatible Specifications. Available online: https://www.thorlabs.com/Images/PDF/Vol18_773.pdf (accessed on 2 September 2017).
23. PicoScope 4824 Data Sheet. Available online: https://www.picotech.com/legacy-document/datasheets/PicoScope4824.en-2.pdf (accessed on 2 September 2017).
24. Intel Core i7-6700K Processor Specifications. Available online: https://ark.intel.com/products/88195/Intel-Core-i7-6700K-Processor-8M-Cache-up-to-4_20-GHz (accessed on 2 September 2017).
25. Geforce GTX 1080 Graphic card Specifications. Available online: https://www.nvidia.com/en-us/geforce/products/10series/geforce-gtx-1080/ (accessed on 2 September 2017).

MDPI
St. Alban-Anlage 66
4052 Basel
Switzerland
Tel. +41 61 683 77 34
Fax +41 61 302 89 18
www.mdpi.com

Electronics Editorial Office
E-mail: electronics@mdpi.com
www.mdpi.com/journal/electronics

www.ingramcontent.com/pod-product-compliance
Lightning Source LLC
LaVergne TN
LVHW070847170726
843515LV00005B/593

* 9 7 8 3 0 3 6 5 0 3 4 4 8 *